Understanding Food Hygiene
And Safety Regulations
(2009/2010 Edition)

Handbook for Food Businesses

FOOD
SOLUTIONS

Food Hygiene and Food Safety Regulations explained

Compiled by
Food Solutions Publishing Ltd

i

Understanding
Food Hygiene and Safety Regulations
(2009/2010 Edition)

Handbook for Food Businesses

First published September 2007 by
Food Solutions Publishing Ltd.

Registered office: 164 Main Road, Goostrey,
Cheshire, CW4 8JP.

Reprinted February 2008 and July 2009

ISBN 978-0-9557466-0-4

A CIP catalogue copy of this Handbook is available from
the British Library

Executive Editor: Bob Salmon.
Assistant Editor: John Golton-Davis.
Sub Editor: Roger Curnock.
Technical Editor: Douglas Brown.
Administration: Jane Davies.
Technical support: Ian Moore.

Printed and bound in England by
Think Digital Print Ltd, Oakham, Rutland

Contents

iii

About the Authors

Bob Salmon, John Golton-Davis and Roger Curnock have at least one thing in common. They have all run their own small food businesses. They each had to learn the changing rules as they went along. So this Handbook is written from the point of view of a small food business owner, not an academic or legalistic expert who may love the whereas, heretofor and similar terms designed to show that "I am cleverer than you".

Bob started and ran a business that pioneered a new product into supermarkets. After selling that business he has worked for some years representing the interests of small food business to the law makers both in London and Brussels. He still works closely with the European organisations and is their expert on the CEN. He sits on three Technical Committees representing some 6 million food SMEs. In Britain he has close links with many trade organisations and professional bodies and is a speaker at British and European conferences. Here he puts across the small business perspective, reminding the Officials that they have a duty to "Think Small First".

Bob and the team would like to thank all those people, too numerous to mention, who have helped with this project. They range from senior officers in the Commission, through FSA officials and consultants, to the many small business people who we asked, "Do you understand this if I write it this way?"

In no place have we interpreted the rules.

All we have done is to put related quotes next to each other so you can see the composite picture without having to refer to too many other documents. None of it would have been possible without the support of the whole team with their technical and administrative skills.

iv

Preface

The food safety laws say there are things you have to do. This handbook tells you **why** the laws are there, **who** has to comply and **what** they say you have to do to comply. It does not tell you **how** to do things as that is a skill specific to your business.

If you want help with the how to comply with food regulations, we suggest you contact your local Environmental Health department, Business Link, the SALSA team, your trade organisation or one of the independent food advisors. There are many useful documents and guides like Safer Food, Better Business and CookSafe, available free from the Food Standards Agency 0845 60 60 667, which give detailed advice.

Much of the material in this book appears in the members' section of Food Solutions website. Here it is amplified and brought together as a reference Handbook. The European Commission publishes on average some 200 regulations, directives or decisions with regard to food each year. This handbook will be modified in subsequent editions. There is also an online version of the Handbook which is updated as necessary. For the full texts of the regulations you are advised to refer to the Food Solutions website www.food-solutions.org. See page 91 for details.

Using the Guide

Texts of Regulations are correct as at the time of publication (July 2009). There may be amendments before the next issue. Food Solutions can inform you of any of these changes, see back cover for details.
The handbook is aimed at anyone within the food sector. Some operators will require basic details (what the law says I have to do). Others may require more detailed information about specific regulations. To accommodate this second group, full texts of the regulations referred to are available on the Food Solutions website and with the online handbook.

To enable readers to find the relevant information quickly, references are written in brackets next to the appropriate text.

For example:

1. *Hygiene of Foodstuffs* (852/2004) refers to the main regulation.
2. *Primary responsibility for food safety rests with the food business operator* (852/2004, chapter1, article1.1a) refers to the specific location of the phrase within the regulations.

We try not to use jargon or unfamiliar words. There are however, words or phrase used in the food regulations, which need to be understood. Glossary of terms (appendix 4) p88.

Part 1: Who and Why? Definitions and Scope

Introduction

Definitions

Risk

Safe Food

Placing on the Market

Chapter 1:

Introduction

"If the law is reasonable, we would respect it"

-Dairyman in Cornwall.

This chapter explains the logic and scope of the food hygiene regulations.

1.1 Overview

Before the introduction of the European Community Food Hygiene Regulations each member state had their own system for regulating the food sector. In the UK there were national regulations, for which each local authority had responsibility for enforcing. Different interpretations were sometimes applied resulting in inconsistencies across the UK.

Pressure was also coming from consumer groups who were insisting that food was safe throughout the EU and that there was a consistent approach to labelling.

The purpose of the legislators has been to satisfy consumer and producer demands for a common approach. They had to pull together all the various regulations in all the member states so that there is one text for everyone. This was difficult as the conditions in the far North are significantly different from those in say Greece or Malta. Coupled with this is the variety of eating habits between the areas of the Union. So the legislators opted for a risk based set of laws putting the onus for compliance on the food business operator rather than the inspectors. In practice the new rules were very similar to the old British ones. **The Editor of this book was involved in the consultations and drafting. He also continues to work on amendments to the Regulations.**

Therefore, at least 95% of food law now comes from Europe. This document seeks to examine why that should be and what that law really is. Extracts from the legislation are printed in italics. The full texts of the legislation are on the Food Solutions website   www.food-solutions.org.

In addition consumer organisations are demanding more information about the foods the public buy in every Member State. Such cross-border agreements are not new; consider international telephone dialling systems and the World Wide Web.

EC Food Hygiene Regulations are in essence very simple

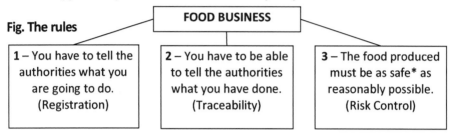

Fig. The rules

| 1 – You have to tell the authorities what you are going to do. (Registration) | 2 – You have to be able to tell the authorities what you have done. (Traceability) | 3 – The food produced must be as safe* as reasonably possible. (Risk Control) |

*Safe food includes labelling, display and advertising

Initial check list.

Food Business operators have to understand the seven key obligations under food safety regulations. Reference is given to the chapter in the Handbook that covers the various obligations.

Safety
Operators shall not place on the market unsafe food or feed. *(Chapters 4, 9, 10, 11 and 12)*

Responsibility
Operators are responsible for the safety of the food and feed which they produce, transport, store or sell. *(Chapters 4 and 5)*

Traceability
Operators shall be able to rapidly identify any supplier or customer. *(Chapter 8)*

Transparency
Operators shall immediately inform their competent authorities (usually your local authority) if they have a reason to believe that their food or feed is not safe. *(Chapter 8)*

Emergency
Operators shall immediately withdraw food or feed from the market if they have reason to believe that it is unsafe. *(Chapter 8)*

Prevention
Operators shall identify and regularly review the critical points in their processes and ensure that controls are applied at these points. (Chapter 7)

Co-operation
Operators shall co-operate with competent authorities in actions taken to reduce risks. (Chapter 7)

These key obligations have been produced by the European Commission Health and Consumer Protection Directorate General. See Website for more information:
http://europa.eu.int/comm/dgs/health_consumer/foodsafety.htm.

This does not presume to be an exhaustive guide to all the regulations. New regulations, directives or decisions regarding food are being discussed every day. Interpretation of regulation is up to the Courts. The national rules usually cover the whole of the UK, but some are specific to the devolved authorities in the parts that make up the UK. We concentrate on those directives and regulations that make up the bulk of food law as applicable throughout Britain and the European Union.

The main Regulations are:

- General Principles of Food Law (EC 178/2002)
- Hygiene of Foodstuffs (EC 852/2004)
- Foods of Animal origin (EC 853/2004)

These are brought into English law by the Food Hygiene (England) Regulations 2006. These UK regulations do not specify what you have to do; that is in the EC legislation. They do define the competent authority (see glossary) for enforcing the provisions and specify the offences and penalties for transgression. Similar regulations apply in the devolved authorities in Scotland and N Ireland. They are amended by the Official Feed and Food Controls (England) Regulations 2007. These give more powers to the Food Standards Agency and to enforcers to charge for any extra inspections.

There are several other regulations, which are pertinent:

- Definition of official controls (EC 854/2004 and 882/2004).
- Transitional arrangements 2074/2005, 2075/2005 and 2076/2005).
- Some amendments in 1662/2006, 1663/2006 and 1665/2006.
- Microbiological criteria for foodstuffs (2073/2005).
- Advertising claims limits (1924/2006).

Some reference is also made to **CEN**, the European standardisation body (see glossary) and **ISO Standards** (see glossary).

As much of the law is rather new, there are few cases proven to set precedents and define interpretation. Also, decisions by English Courts may not be upheld in the European system. It is to the credit of the legislators that these regulations have not needed significant amendments since they were first written.

Knowing the rules applicable to your business is part of taking reasonable precautions. Ignorance is not an excuse.

1.2 Why the law?

The laws are there firstly to ensure the free movement of goods between member states.
"The free movement of food and feed within the Community can be achieved only if food and feed safety requirements do not differ significantly from Member State to Member State" (178/2002 recital 3).

This means that the same rules should apply in each member state. Thus consumers and traders can buy and sell without worrying about local restrictions.

The second reason given is that food and feed should be safe in every country.
"A high level of protection of human life and health should be assured in the pursuit of Community policies. The Community has chosen a high level of health protection as appropriate in the development of food law, which it applies in a non-discriminatory manner whether food or feed is traded on the internal market or internationally" (178/2002 recitals 2 and 8).

This means that when you are on holiday in Spain or any other EU country you can be sure that the food you are eating is produced to the same rules as in England. Similarly the labelling rules are the same. It means that every producer and trader has to apply the same safety standards. It gives confidence to the eaters and a level field for the suppliers.

Practical example:

In the UK you always buy free range eggs. European Regulations state, with a few exceptions, that eggs must be marked giving you that information. The first number that appears tells you whether the egg is Organic (0); Free range (1); Barn (2); Caged bird (3). So a free range egg bought in England will be printed:

1 UK followed by a code for the producing farm.

You are on holiday in France and you want free range eggs.
Simple: eggs will be printed:

1 FR followed by a code for the producing farm.

This means that without knowing a word of French you can be sure that the eggs you buy meet your requirements (see chapter 13.3 for full details).

A further objective of food law is the protection of the interests of consumers.

It should allow consumers to make informed choices. It shall aim at the prevention of:

1. Fraudulent or deceptive practices.
2. The adulteration of food.
3. Any other practices, which may mislead the consumer (article 8 of 78/2002).

This statement in the Regulation is the basis upon which many of the detailed rules are developed.

The EC regulations have **"recitals"** (see glossary) at the beginning, which explain the reasons for things, then **"articles"** (see glossary) within the regulation, which are the mandatory statements (see glossary).

There are seven basic principles that all EC food law is based on. These are:

1. Protection of human life and health.
2. Protection of consumers' interests.
3. Fair practices in the food trade.
4. Free movement of food and feed throughout the EU.
5. Protection of animal health and welfare.
6. Protection of plant health.
7. Protection of the environment.

There is no indication as to which of these principles takes precedence (178/2002 article 5).

These 3 objectives are why the regulations have been made. The question now is **why** you should be aware of them and **why** you should take all necessary steps to comply.

The process of introducing new food regulations.

The process from the original discussions to the food regulations being introduced follows a set pattern. The following flow chart shows the various stages for the introduction of a new or amended food regulation, from inception to the introduction in the UK. As you can see this is a long process with many opportunities to influence the final outcome. The problem for many smaller businesses is that the first they hear of a new regulation is when it has been, or is about to be introduced. By then it is very difficult to change the outcome.

Fig. The stages for introducing food regulations

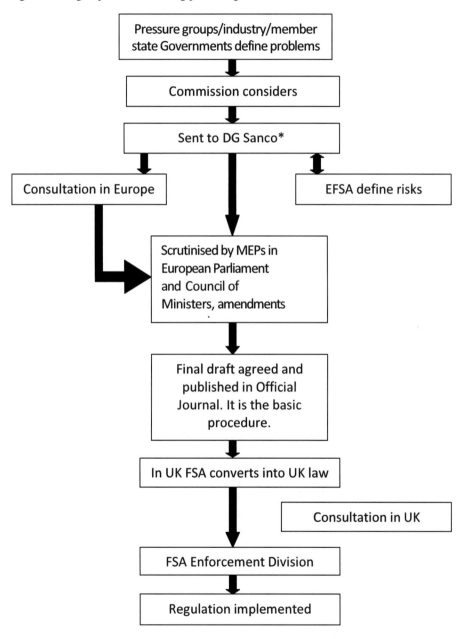

* **DG Sanco - see glossary**

The writers of regulations are skilled and take into account input from the food industry. This case study illustrates this.

Case study: Small abattoir

An inspection at a small abattoir resulted in the owner being informed that he would be closed down as the regulations stated that there had to be separate areas for each process.

The writers of the regulations were aware that small abattoirs existed so the regulation states that separation can be either in space or time.

The owner of the abattoir was unaware of this regulation, and the inspector had overlooked it. This could have resulted in the closure of the premises causing a major problem for the farmers who depended on this facility.

Learning point

When you learn to drive, you are instructed how to drive. You must also know what the traffic law says you have to do. If you run a food business you need to know how to do the job. You also need to know what you have to do to comply with the law.

1.3 Why you?

If, and we hope it never happens, you had a claim against you, which is increasingly likely in this age, where no win no fee solicitors are encouraging the public to sue, you must have a suitable defence. It is no good saying that you did not know or that no one told you.

The only admissible defence is one of due diligence.

This is spelled out in the Food Hygiene (England) Regulations (2006 SI 2006 No 14). Similar documents exist for Scotland and Northern Ireland, which define the powers of officials and the offences.

Clause 11 says; *In any proceedings for an offence under these regulations, it shall … be a defence for the accused to prove that he took all reasonable precautions and exercised all due diligence to avoid the commission of the offence by himself or by a person under his control.*

This is known as the due diligence defence.

Thus taking all reasonable precautions is like paying the insurance premium that your business is not going to be closed down by any authorised officer.

The key word here is "prove". To do this you must be able to demonstrate effectively that you have done what you say you have done.

This means keeping records of things (like training or cleaning) and finding out what the rules really are.

Practical example

A new kitchen assistant is taken on. For the first week he is working alongside an experienced operator who teaches him the procedures. A month later he makes a mistake and the inspector questions whether he was taught properly. Without anything written down the business owner could be liable.

Practical example 2

A new kitchen assistant is taken on. She goes through a formal recorded induction covering all aspects of the business and food regulations and is constantly monitored to test for understanding. All of this is fully documented and recorded. A month later she makes a mistake, the inspector can see that full training was given. Hence, the owner could offer a defence.

1.4 General Scope

This General Food Law EC178/2002 provides a framework for all the other food laws and is applicable to all sectors of the food industry. Thus it is called a horizontal regulation and forms the basis for this guide. There are another set of regulations known as vertical regulations usually these are specific to a product or class of products and means that they have to comply with additional regulations. The Honey, Jams and Egg Regulations are examples. This handbook does not cover vertical regulations in this edition, as anyone affected should be aware of the extra requirements.

Fig. Horizontal and vertical regulations

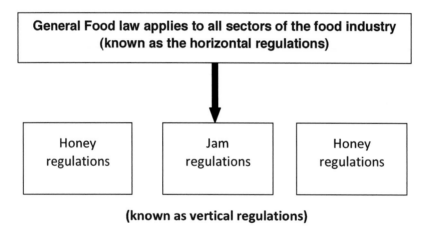

General Food law applies to all sectors of the food industry (known as the horizontal regulations)

| Honey regulations | Jam regulations | Honey regulations |

(known as vertical regulations)

One of the new provisions within these documents is that you, the Food Business Operator (FBO), are responsible.

It is no defence, if something were to go wrong, to say that you had just been inspected, accredited or even approved.

The rules are quite clear.

> *Primary responsibility for food safety rests with the food business operator.*
> **(852/2004 chapter 1, article 1.1.a)**

That article goes on to require food safety throughout the food chain, from primary production – that is farmers or hunters – to the final consumer. It adds that microbiological and chemical standards are to be introduced to ensure the minimum of risks.

The full provisions of this regulation do not apply to:

- *The primary production for private domestic use.*
- *The domestic preparation, handling and storage of food of private domestic consumption.*
- *The direct supply, by the producer, of small quantities of primary products to the final consumer or to local retail establishments directly supplying the final consumer.*

This is amplified in the EC guidance to include farm gate sales, local markets and local restaurants. It also covers people who collect wild mushrooms and berries. Local is sales within your own county plus the greater of either neighbouring county or counties or 30 miles/50 Km from the boundary of your county.

Chapter 2:
Definitions

"I am a bear of very little brain and long words confuse me" –A. A. Milne.

These are the official definitions.

In common with all the EC regulations there are certain definitions laid down at the start. We need to do the same here so we know what we are talking about. The most important one is the definition of food. This definition is being read across into all other regulations and slowly into UK laws. Safe food and Placing on the Market in chapters 4 and 5 are also critical definitions. See also appendix 2 for a list of abbreviations.

2.1 Food

Food (or 'foodstuff') means any substance or product, whether processed, partially processed or unprocessed, intended to be, or reasonably expected to be ingested by humans.
'Food' includes drink, chewing gum and any substance, including water, intentionally incorporated into the food during its manufacture, preparation or treatment. It includes water after the point of compliance as defined in Article 6 of Directive 98/83/EC and without prejudice to the requirements of Directives 80/778/EEC and 98/83/EC.
Food' shall not include:
(a) Feed.
(b) Live animals unless they are prepared for placing on the market for human consumption.
(c) Plants prior to harvesting.
(d) Medicinal products within the meaning of Council Directives 65/65/EEC and 92/73/EEC.
(e) Cosmetics within the meaning of Council Directive 76/ 768/EEC .
(f) Tobacco and tobacco products within the meaning of Council Directive 89/622/EEC.
(g) Narcotic or psychotropic substances within the meaning of the United Nations Single Convention on Narcotic Drugs, 1961, and the

United Nations Convention on Psychotropic Substances, 1971.
(h) Residues and contaminants.

So food is almost everything that could be eaten by anyone

2.2 Feed

Feed (or 'feedingstuff') means any substance or product, including additives, whether processed, partially processed or unprocessed, intended to be used for oral feeding to animals.

2.3 Food business

Food Business means any undertaking, whether for profit or not and whether public or private, carrying out any of the activities related to any stage of production, processing and distribution of food.

2.4 Food Business Operator (FBO)

Food Business Operator means the natural or legal person responsible for ensuring that the requirements of food law are met within the food business under their control (178/2002 articles 2 and 3).

This last definition is all-embracing in that it covers the largest supermarket as well as the school fete. The only exception is cooking for your family in your own home. *It shall not apply to primary production for private domestic use or to the domestic preparation, handling or storage of food for private domestic consumption* (178/2002 article 1). This regulation specifies traceability (see chapter 8). We shall see later that this definition is not always applied.

Traceability is demanded of everyone.

HACCP (see chapter 7) is not required of some very low risk activities. Some labelling rules (see chapter 13) only apply to pre-packed goods.

2.5 Charity functions

Operations such as the occasional handling, preparation, storage and serving of food by private persons at events such as church, school or village fairs are not covered by the full scope of the hygiene regulations. This is made clear in recital 9 of 852/2004, which says *"Community rules should only apply to undertakings, the concept of which implies a certain continuity of activities and a certain degree of organisation".* The term *"undertaking"* is integrated into the definition of a food business. *Somebody who handles,* prepares, stores or serves food occasionally and on a small scale (e.g. a church, school or village fair and other situations such as organised charities comprising individual volunteers where food is served occasionally) cannot be considered as an undertaking and is therefore not subject to the requirements of Community hygiene legislation. (EC guidance on 852/2004 dated December 2005). They do come under the 178/2002 rules for safe food and traceability.

It is only HACCP (see chapter 7) and registration (see chapter 6) that charity functions may not be subject to.

2.6 Risk analysis
Risk analysis is three things; assessment, management and communication. These are defined as:
'Risk assessment' means a scientifically based process consisting of four steps: hazard identification, hazard characterisation ,exposure assessment and risk characterisation.
'Risk management' means the process, distinct from risk assessment, of weighing policy alternatives in consultation with interested parties, considering risk assessment and other legitimate factors, and, if need be, selecting appropriate prevention and control options.
'Risk communication' means the interactive exchange of information and opinions throughout the risk analysis process as regards hazards and risks, risk-related factors and risk perceptions, among risk assessors, risk managers, consumers, feed and food businesses, the academic community and other interested parties, including the explanation of risk assessment findings and the basis of risk management decisions.

This means:
Is there a significant risk?
Can I control it?
Do I need to tell someone about it?

2.7 Hazards
Hazard means a biological, chemical or physical agent in food, or the condition of food or feed, with the potential to cause an adverse health effect (all these definitions are from 178/2002 article 3).

Chapter 3:
Risk

"The concept of zero risk is unobtainable. The question is, what level of risk might be safe enough?" **- Sir John Krebbs as CEO of the FSA.**

There is risk in everything. Your business skill is in managing and controlling all of the risks from food hygiene to the accounts.

There is no such thing as the total absence of risk. Risks include those to yourself, your staff and your customers both from your products and from your premises/equipment. Risks are not just about being found out or invoices not being paid. Nor are they just things to insure against. The reality is that most food operators carry out risk assessment as part of their own best practice – it does not make commercial sense to poison their customers.

As the laws could not be specific, the writers opted for risk analysis. *Where food law is aimed at the reduction, elimination or avoidance of a risk to health, the three interconnected components of risk analysis — risk assessment, risk management, and risk communication — provide a systematic methodology for the determination of effective, proportionate and targeted measures or other actions to protect health* (178/2002 recital 17).

Risk is defined in the regulation as: *'risk' means a function of the probability of an adverse health effect and the severity of that effect, consequential to a hazard* (178/2002 article 3). Also see under definitions in chapter 2.

This means: Is something likely to happen and, if it did, how serious would be the result?

The General Principles Regulation EC 178/2002 then set up the European Food Safety Authority (EFSA) to advise the politicians and legal writers. The Authority shall provide scientific advice and scientific and technical support for the Community's legislation and policies in all fields, which have a direct or indirect impact on food and feed safety. It shall provide independent information on all matters within these fields and communicate on risks. Also to provide the Community institutions and the Member States with the best possible scientific opinions in all cases provided for by Community legislation and on any question within its mission (178/2002 articles 22 and 23). Article 23 goes on to list the twelve tasks of the Authority. This Authority, known as EFSA, is now based in Palma in Italy. It uses the scientific and academic resources of all the member states to generate its opinions and advice.

There is no such thing as the total absence of risk.

It is your duty to control them to acceptable levels:

- Identify the risks; from whatever cause.
- Assess the likelihood of them happening.
- Devise and implement control strategies (this is almost HACCP).

See HACCP in chapter 7 – this is the system for controlling risks to your product – and the Health and Safety provisions in chapter 12.

3.1 Risk assessment

The official definition of risk assessment is given in chapter 2.6. The four steps need consideration:

1. Identify the hazards such as biological (bacteria and spoilage), chemical (residues of cleaners or even allergens) or physical (bits of glass, metal or plastic).
2. Examine whether any of these could cause a problem. This is called hazard characterisation.
3. Look at the exposure. Is it likely that the amounts are significant?
4. Decide whether, in the light of the first three answers, there is a significant risk. It is the combination of the answers to these questions that defines a hazard – highly likely/big problem – to a possibility – unlikely/minimal effect.

Practical example of a non-food risk assessment

Consider crossing a road. The likelihood of a problem is whether you can see clearly, how wide the road is, how quickly you can walk, are you alert? The scale of the problem is defined by how fast cars are coming, size of approaching vehicles etc. Your decision to cross depends on these factors.

Doing a risk assessment is to look at the questions in your business with reference to the way you operate.

Case study: Blue Steak

A restaurateur was told that he could no longer provide blue steaks for his customers.

The regulations say: *Primary responsibility for food safety rests with the food business operator* (852/2004 chapter 1, article 1.1.a).

This means that, if the restaurateur has satisfied himself that the product is safe and he has assessed any possible risks in his cooking method, it is his decision whether or not to serve blue steak.

To demonstrate due diligence - a record of the risk assessment process carried out would be appropriate.

Other typical cases where a food business operator should have carried out a risk assessment:

- The hand washing basin was used for straining vegetables.
- The chopping boards used for raw and cooked meats were stacked on top of each other when not in use.
- "We did training in the 1980s. Has anything changed?"
- "I didn't know the cleaning chemical had to be in contact with the surface for so long".
- In the case study outlined in chapter 14 the school assessed the probability of allergen contamination as possible and the severity as fatal. They therefore decided on a complete ban on nuts throughout the school.

If you want help with how to carry out risk assessments we suggest you contact your local Environmental Health department, Business Link, the SALSA team, your trade organisation guide to good practice or one the independent food advisors.

Food Solutions has produced several detailed articles on the subject of risk. These are available to subscribers on the website.

Chapter 4:
Safe Food

No food is universally safe. You can die from an overdose of water. You can become ill through a lack of salt. It is not the intention of this Guide to define a good diet. It is the intention to draw your attention to what the law actually says. In this respect the law is quite clear.

Food shall not be placed on the market, if it is unsafe (178/2002 article 14).

This definition is important, as we will see later that a use by date shows when a food is likely to become unsafe. So it is an offence to sell foods after the use by date. The best before date only indicates that the product may not be in the best condition after that day (see chapter 13.2).

The important point here is to note that it is "the intended use" that a product is assessed on, also whether the product is "actually" placed on the market. It is not an offence to have on your premises a product that is patently unfit but in your waste bin.

The full definition is as follows:

1. ***Food shall not be placed on the market*** - *if it is unsafe.*

2. *Food shall be deemed to be unsafe - if it is considered to be:*
   a) *injurious to health;*
   b) *unfit for human consumption.*

3. ***In determining whether any food is unsafe*** - *regard shall be had:*
   a) *to the normal conditions of use of the food by the consumer and at each stage of production, processing and distribution;*
   b) *to the information provided to the consumer, including information on the label, or other information generally available to the consumer concerning the avoidance of specific adverse health effects from a particular food or category of foods.*

4. **In determining whether any food is injurious to health** - *regard shall be had:*
   a) *not only to the probable immediate and/or short-term and/or long-term effects of that food on the health of a person consuming it, but also on subsequent generations;*
   b) *to the probable cumulative toxic effects;*
   c) *to the particular health sensitivities of a specific category of consumers where the food is intended for that category of consumers.*

> **Key point: foods containing allergens not shown on the label are unsafe.**

5. **In determining whether any food is unfit for human consumption** - *regard shall be had to whether the food is unacceptable for human consumption according to its intended use, for reasons of contamination, whether by extraneous matter or otherwise, or through putrefaction, deterioration or decay.*

6. **Where any food which is unsafe is part of a batch, lot or consignment of food of the same class or description** - *it shall be presumed that all the food in that batch, lot or consignment is also unsafe, unless following a detailed assessment there is no evidence that the rest of the batch, lot or consignment is unsafe.*

7. **Food that complies with specific Community provisions governing food safety** - *shall be deemed to be safe insofar as the aspects covered by the specific Community provisions are concerned.*

8. **Conformity of a food with specific provisions applicable to that food** - *shall not bar the competent authorities from taking appropriate measures to impose restrictions on it being placed on the market or to require its withdrawal from the market where there are reasons to suspect that, despite such conformity, the food is unsafe* (178/2002 article 14).

The inference here is that raw chicken, which may be slightly contaminated, is deemed safe whereas chocolate with the same degree of contamination would be unsafe. That is because the presumption is that the chicken would be cooked and the chocolate would not. In this context cooking instructions should be included on the labels of pre-packed products.

Food is also considered unsafe if it contains any of the known allergens. Allergens are dealt with in detail in chapter 14. Clause 4c above implies that foods containing either allergens or other ingredients, which may be unsuitable for certain classes of consumer, must be labelled accordingly. This applies particularly to foods, which may be fed to infants, pregnant women or aged persons. Specific rules have been drawn up for baby foods and compounds aimed at certain classes, for instance sports drinks.

Chapter 5:
Placing on the market

This is defined as: *the holding of food or feed for the purpose of sale, including offering for sale or any other form of transfer, whether free of charge or not, and the sale, distribution and other forms of transfer themselves* (178/2002 article 3).

That also means internet and mail order sales.

This is amplified in the UK Food Hygiene Regulations 2005. *Any food commonly used for human consumption shall, if placed on the market or offered, exposed or kept for placing on the market, be presumed, until the contrary is proved, to have been placed on the market or, as the case may be, to have been or to be intended for placing on the market for human consumption.*

Any food commonly used for human consumption, which is found on premises used for the preparation, storage, or placing on the market of that food and any article or substance commonly used in the manufacture of food for human consumption, which is found on premises used for the preparation, storage or placing on the market of that food, shall be presumed, until the contrary is proved, to be intended for placing on the market, or for manufacturing food for placing on the market, for human consumption.

Any article or substance capable of being used in the composition or preparation of any food commonly used for human consumption, which is found on premises on which that food is prepared, shall, until the contrary is proved, be presumed to be included for such use.

This definition does not apply to the requirement to inform the authorities when your product is not in compliance and so has to be withdrawn. Here the regulation says: *has left the immediate control* (178/2002 article 19.1).

The rationale here is that you may be able to take corrective action to bring the food into compliance or scrap it. It would be silly to have to tell the authorities every time you burned the cakes.

This means that, if you have a bottle of outdated sauce in the back of the cupboard, you have committed an offence.

21

Part 2: What and Where? Specific Rules

Registration

HACCP

Traceability

Premises

Equipment

Personal Hygiene

Health and Safety

Labelling

Allergens

Temperature Control Requirements

Training

Chapter 6:

Registration

EC Food Hygiene Regulations are in essence very simple:

- You have to tell the authorities what you are going to do.
- You have to be able to tell local authorities what you have done.
- The food produced must be as safe as reasonably possible.

The first of these is Registration. The law says that:

> **All food operators are required to register so that that the authorities know where you are and what sort of business you have.**

The regulation is:

Every Food Business Operator (FBO) shall notify the appropriate competent authority, in the manner that the latter requires, of each establishment under its control that carries out any of the stages of production, processing and distribution of food, with a view to the registration of each such establishment. FBOs shall also ensure that the competent authority always has up-to-date information on establishments, including by notifying any significant change in activities and any closure of an existing establishment (852/2004 article 6).

This means that your local Council Environmental Health Department must be told what you are doing.

Registration forms are available from your Local Authority.

In the UK the Competent Authority is the Food Standards Agency, except where it has delegated competencies to other bodies. This is specified in the UK Food Hygiene Regulations. Without going into too much detail, most of the general legislation is enforced by the468 Local

Councils through their Environmental Health Practitioners (EHPs) and officers. This applies to almost all the requirements of the EC Hygiene Regulation 852/2004, which demands registration and HACCP.

The Local Councils also have Trading Standards Officers, who look at labelling, weights and measures and other details.

The EC regulation 853/2004 supplements 852/2004 with special rules for products of animal origin.

- Some of this is enforced by the Meat Hygiene Service, which is an executive arm of the Food Standards Agency.
- Some slaughterhouse operations are overseen by the Official Veterinary Service, which is also affiliated to the FSA.
- Imports and some exports are checked by the Port Health Authorities.
- Primary production is mainly the province of the Department of the Environment and Rural Affairs (DEFRA) who also, just as an anomaly, supervise egg marking.

Most of this is defined in the Food Hygiene (England) Regulations 2006 clauses 4 and 5.

If you think that this is complex, the systems used in some of the other member states are far more intricate.

Some businesses dealing with products of animal origin also have to be approved. These are mainly abattoirs and meat cutting plants where the official stamps are put onto meats. The regulation says: *....shall not operate unless the competent authority has ... granted the establishment approval to operate following an on-site visit* (853/2004 Article 4). Examples of such establishments include slaughterhouses (except for small quantities of poultry or rabbits for direct sale), cutting plants, game handlers (see glossary), milk processing (heat treating, cheese-making etc.), egg processors, renderers (see glossary) and cold stores where bulks are broken down. There is a list in annex IV of the EC guide to 853/2004.

Registration is through your Local Council EHP. You can get a form to fill in from them.

Does your local authority know what you are doing? **Yes or No**	
When did you last notify them in writing? **Date**	

REGISTRATION

Chapter 7:
Hazard Analysis & Critical Control Points (HACCP)

This chapter looks at controlling the risks.

Background

When the Americans wanted to send men into space they were concerned that the food they sent with them should be safe. They did not want the men to suffer tummy upsets in weightlessness, as that could have been unfortunate. Up till then most food testing had been done on the finished product.

Fig. Showing difference between the tradition approach and the HACCP approach.

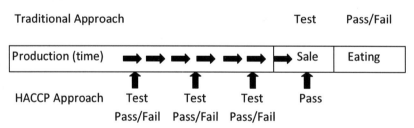

The result of tests from the traditional approach might not be available until the food had been distributed and eaten.

For astronauts that would have been too late. So they, with consultants, devised a system for their food suppliers where testing was done at each stage of production. That meant that the final product could be guaranteed to be OK. They thought that checks at each critical point along the manufacturing process would be a good idea. The system was so effective that it has now become part of international food regulations.

The system was introduced in Britain in the now revoked 1995 Food Safety Regulation, but they only demanded the first five steps of the principles:

1. Identifying any hazards that must be prevented.
2. Identifying the critical control points.
3. Establishing critical limits at critical control points.
4. Establishing and implementing effective monitoring.
5. Establishing corrective actions.
6. **Have written procedures.**
7. **Keep records.**

They did not require all businesses to keep records of the checks they did. This was because the EC was drafting new rules and everyone expected that the Commission would demand all seven steps. In the event, the Commission Regulation left open various options for smaller and less risky businesses to design simplified systems.

This meant that there has been some confusion in Britain as many people thought the EC would demand the full system whereas in fact they did not. To rectify this the FSA has designed some guides. In England it is called Safer Food, Better Business. In Scotland it is called CookSafe. The Northern Ireland version is Safe Catering. All these can be obtained free from the FSA. Just to complicate matters there are several versions specific to catering and to retailers. These may not be suitable for all types of food business. We will look first at what the law actually says and then at what that means.

Many books have been written on HACCP and it has been incorporated into international standards. The standard definitions are contained in the Codex Alimentarius version. This was put together by the World Health Organisation Food Standards Programme. It has been developed and incorporated into the international standard ISO 22,000. This standard suggests a full time HACCP co-ordinator with a committee to oversee implementation. There are some 16 duties of "top management" and at least 67 separate records to be kept. This may be suitable for some small
catering and retail businesses.

A simpler and more user friendly system is defined in the **SALSA** (see glossary) scheme designed more for small businesses. SALSA is Safe And Local Supplier Approval and is sponsored by the British Retail

Consortium, The British Hospitality Association, the Food and Drink Federation and the National Farmers union.

7.1 Basic requirements

The law, defined in EC 852/2004, which came into force on 1 January 2006, demands two things:

First - is what is called the "**prerequisite requirements**" or *specific hygiene measures* (article 4). These are mainly common sense and are to do with general hygiene. They include:

Safe handling	Sanitation	Training
Raw materials	Sampling and analysis	Water quality
Waste management	Cold chain maintenance	Traceability
Pest control		

The law is quite specific in article 4 of 852/2004 where it says; *food business operators shall, as appropriate, adopt the following specific hygiene measures...* it also talks about; *meeting targets set to achieve the objectives of this Regulation.*

The Regulation sets targets. It is up to you how you get there.

It then specifies all the requirements for premises in the annexes (see chapter 9). Cold chain temperatures are defined in Schedule 4 of the Food Hygiene (England) (no2) Regulations 2005 and the Annex to EC 853/2004.

Second - is the demand: *to put in place, implement and maintain a permanent procedure or procedures based on the HACCP principles* (article 5).

Guidance on safe handling and good practice is shown in Safer Food, Better Business (SFBB) and its cousins. These are very useful documents as they define the things to be done on a national basis. Time was when each of the 468 Local Authorities each had its own recommendations.

Hazard Analysis and Critical Control Points (HACCP) comes in after these pre-requisites have been done. HACCP does not make food safe. It is a management system to make sure that you and your staff have taken all reasonable precautions. However, the full system and an abridged system requirement for low risk activities are defined in the EC guidance to HACCP.

> **The law does not say that you must have a HACCP system, but a permanent procedure based on the principles of HACCP.**

Where the prerequisites, whether or not supplemented with guides to good practice, achieve the objective of controlling the hazards in food, it should be considered, based on the principle of proportionality, that the obligations laid down under the food hygiene rules have been met and that there is no need to proceed with the obligation to put in place, implement and maintain a permanent procedure based on the HACCP principles. (EC guide to HACCP)

These principles are laid out in the regulation 852/2004 article 5.

1. *Food business operators shall put in place, implement and maintain a permanent procedure or procedures based on the HACCP principles.*

2. *The HACCP principles referred to in paragraph 1 consist of the following:*

    1) *identifying any hazards that must be prevented, eliminated or reduced to acceptable levels;*

    2) *identifying the critical control points at the step or steps at which control is essential to prevent or eliminate a hazard or to reduce it to acceptable levels;*

    3) *establishing critical limits at critical control points which separate acceptability from unacceptability for the prevention, elimination or reduction of identified hazards;*

    4) *establishing and implementing effective monitoring procedures at critical control points;*

    5) *establishing corrective actions when monitoring indicates that a critical control point is not under control;*

    6) *establishing procedures, which shall be carried out regularly, to verify that the measures outlined in subparagraphs (1) to (5) are working effectively;*

7) establishing documents and records commensurate with the nature and size of the food business to demonstrate the effective application of the measures outlined in subparagraphs (1) to (6). When any modification is made in the product, process, or any step, food business operators shall review the procedure and make the necessary changes to it (852/2004 article 5(1)).

The Regulation also gives three more duties:
Food business operators shall:
1. *Provide the competent authority with evidence of their compliance with paragraph 1 in the manner that the competent authority requires, taking account of the nature and size of the food business;*
2. *Ensure that any documents describing the procedures developed in accordance with this Article are up-to-date at all times;*
3. *Retain any other documents and records for an appropriate period.*

Before we look at these seven stages in detail we should look carefully at the words **"nature and size of the business"**.

According to the authors of the legislation they do not mean just size but they are a measure of risk.

- A business selling ready to eat foods to many people has high risk.
- A business retailing pre-packed product to a few may have low risk.

The EC guidance on HACCP is in two parts. The first details all the things that a high-risk enterprise has to do. The second introduces many derogations, mainly in the number and nature of the records to be kept. A "**derogation**" (see glossary) allows defined activities to be exempt from that part of regulation. SFBB and its cousins steer a middle road and recommend reporting by exception rather than by rule.

The first stage is identifying the hazards. These may be:
- Biological, in the form of contamination with bacteria, yeasts or moulds (gone off).

- Chemical, in the form of cleaning fluids, pesticides or lubricants.
- Physical, in the form of broken plastic or glass.

Contamination may also be from allergenic ingredients that have been introduced by mistake. All these make the product unsafe (see chapter 4 on safe foods). Now assess the risks from all these possibilities and how they may be reduced or eliminated.

Second: look at the points in your process that are critical.
These may be:
- The state of materials coming in to you.
- Your cooking temperatures.
- Storage, packaging etc.

What are the acceptable levels? Many of these are dealt with under the microbiological criteria regulations and there are limits laid down and rules for the frequency of testing. Other legal limits are on contamination of residues, heavy metals etc. and of course temperatures.

Most acceptable levels are determined by visual inspection. The EC guidance to HACCP says: *ensure that appropriate control measures are effectively designed and implemented. In particular, if a hazard has been identified at a step where control is necessary for product safety and no control measure exists at that step, then the product or process should be modified at that step or earlier to include a control measure.*

Third: the critical limits that you can accept. Do not forget that some bacteria can double every 20 minutes in the right conditions. Critical limits *correspond to the extreme values acceptable with regard to product safety. They are set for observable or measurable parameters, which can demonstrate that the critical point is under control* (EC guide to HACCP).

Fourth: establishing the effective monitoring - may sound difficult, but here again such basic things as cooking and storage temperatures make a difference.

An essential part of HACCP is a programme of observations or measurements performed at each critical point to ensure compliance with specified critical limits. You must have a procedure that makes it clear as to who performs the monitoring, when and how it is done.

Fifth: the corrective actions that you need to take - are largely defined by experience and training.

Sixth: you should have a written procedure - of who is responsible for corrective action, what that action should be and a record of the action taken.

Seventh: records need to be kept. This is both for your sake to demonstrate due diligence and so that you can prove to any authority that the checks and actions have been carried out. The guidance says that documents should be signed by a responsible reviewing official of the company.

The regulation also says that you must be able to demonstrate that you have done all the right things in a way that is acceptable.

> *Food business operators shall provide the competent authority with evidence of their compliance in the manner that the competent authority requires, taking account of the nature and size of the food business* **(852/2004 article 5.3).**

The acceptable manner in the UK is generally as set out in SFBB and its cousins. Here reporting is done more by exception than by rule. Businesses are encouraged to keep a diary that the checks have been carried out and only record when things go wrong and, of course, what action was taken to correct any deviations.

The final stage in the HACCP process is to review your system regularly. In conjunction with the UK competent authorities, Food Solutions has drawn up a checklist to help with the review process. It is not the HACCP system. It is a memory aid to stimulate you to ask yourself some of the questions that may need to be asked. Every question may not be applicable to your operation.

Checklist

The items highlighted in **bold** are key checks. Under the food regulations the only admissible defence is one of due diligence (see chapter 1.3). The key word here is "prove". To do this you must be able to demonstrate effectively that you have done what you say you have done. This means keeping records and having in place procedures for these crucial issues.

The references are to chapters within this book and to the FSA Safer Food, Better Business (SFBB).

	Review Date	Review Date	Review Date
Have you systematically thought through and carefully reviewed all the key food safety issues relating to your business? Have any risks changed?			
Have you worked out what needs to be done to avoid things going wrong and have you written down your resultant food safety procedures?			
Do you verify the operation of food safety checks and do you keep a simple record of the checks or at least the fact that the checks have been done? See SFBB opening checks			
Have all your staff received appropriate training for your business? See training in chapter 16			
Do you have a system for withdrawing or recalling unsafe product? See traceability for "unsafe" in chapters 8 and 4			
Do you have a system for traceability of products? See traceability in chapter 8			
If something went wrong, could you say where you got all your supplies from and what you have sold to other businesses?			
Do you use food grade equipment/packaging?			
Do you know what allergens you may have/use? See allergens in chapter 14			
Are you confident with your suppliers? Do your staff know what to do if there is any doubt?			
Do you check the temperature of incoming goods and note down the fact that the check has been done?			

	Review Date	Review Date	Review Date
Do you know the temperatures at which different foods should be stored? See chapter 15			
Can you demonstrate that your fridge and freezer are keeping to the right temperatures?			
Do your staff really know what cross-contamination is and how to prevent it? It applies to allergens as well as contamination. See allergens in chapter 14			
Can you demonstrate that the way you handle and prepare food will prevent cross-contamination?			
Do you use separate equipment for raw and cooked foods? See chopping boards in guidance notes on Food Solutions Website.			
Can you demonstrate that those foods are properly separated?			
Do your staff know what temperature will kill most bacteria?			
Do you have written procedures to define cooking/reheating times?			
Can you demonstrate that your temperatures are accurate?			
Do you have a reliable system for establishing the "shelf life" of foods once they have been cooked?			
Are you aware of the dangers that "may" happen when cooking/reheating in a microwave? See Microwave Ovens in chapter 10.3			
Can you demonstrate that your cooling times are right?			
Do you comply with the Fire Safety Order as from October 2006? Every business has to. See Fire Regulations in chapter 9.5			
Do you comply with the Gas rules? See Gas Regulations on Food Solutions Website			
Do you have adequate ventilation? See Premises in chapter 9			
Are the kitchen walls/floors/ceilings in good repair and washable?			
Do you have a hand washing basin in a convenient place?			
Do you have adequate hand drying facilities?			
Do your staff have suitable protective clothing?			

	Review Date	Review Date	Review Date
Do you have adequate storage for waste and in the right place? See www.netreg.gov.uk for the environmental rules.			
Are your hygiene rules written down so your staff can refresh their memories?			
Do you have a first aid kit properly stocked?			
Do you have a system for staff to tell you about any illness?			
Do you ask new staff to complete an employee medical questionnaire to ensure that they are fit to work in a food preparation area?			
Do you keep staff sickness records?			
Do you have a cleaning routine?			
Could you demonstrate that your staff keep to it?			
Are you sure you use the right chemicals at the right temperatures and for the right times at the right strength for effective cleaning?			
Do you store your cleaning materials in a safe area away from foods?			
Do you dispose of waste in the right way? See Premises in chapter 9.			
Are your premises properly pest proof?			
Do you use a pest control contractor?			
Can you demonstrate that you regularly check for any contamination?			
Can you demonstrate that you have taken the right corrective action whenever something has gone wrong?			
Can you demonstrate that you have identified any critical points in your procedures and have set safe limits for them?			
Can you demonstrate that your staff are doing all the hygiene things and checks necessary even when you are not there?			
Are you confident about Cross-contamination; Cleaning; Chilling; Cooking? Read SFBB again			
If you are confident, then sign and date here to say you have considered all the questions.			

The questions above may remind you of some of these requirements. The EC guide to HACCP says:-

Once these are done it may seem that all the hazards can be controlled in an operation where there is no preparation, manufacture or processing of food. There is then no further need to develop the HACCP stages. Such operations may include market stalls, coffee shops, small retailers and some transporters.

In an operation where guides to good practice are followed for food processing and preparation there would be no need to further develop HACCP. Such operations could include restaurants, caterers, pizza shops, bakers and butchers.

There is a requirement that all staff should be trained in HACCP procedures so that they know what they have to do, when and why. The regulation says: *Food business operators are to ensure that food handlers are supervised and instructed and/or trained in food hygiene matters commensurate with their work activity and that those responsible for the development and maintenance of the procedure referred to in Article 5(1) of this Regulation or for the operation of relevant guides have received adequate training in the application of the HACCP principles* (852/2004 annex II chapter 12). The procedure referred to in article 5(1) is the seven stages of HACCP reproduced above.

The practical details of an acceptable HACCP system are well shown in the Safer Food, Better Business packs and their cousins CookSafe and Safe Catering (Scotland and N Ireland) with photo guides to good practice and possible risks. Rather than reproduce all that information here we would recommend you to obtain free copies from your local Environmental Health office or free from the FSA on 0845 60 60 667.

Examples of issues that should have been addressed when carrying out HACCP

- "Well, I keep records of temperatures and things. Isn't that enough?"
- Overalls and protective clothing worn when coming to work on the bus. These were then washed at home together with the children's sports kit.
- The cleaning company recorded as being responsible had gone out of business two years before.
- Colour coded tongs for raw and cooked meats were stacked on top of each other when not in use.

If you want help with how to carry out HACCP, we suggest you contact your local Environmental Health department, Business Link, the SALSA team, your trade organisation guide to good practice or one the independent food advisors.

Food Solutions has produced several detailed articles on the subject of HACCP. For details of how to access these see page 91.

When did you last review your critical points and their limits? *Date*	
Can you demonstrate that you are controlling risks? *Yes or No*	

H
A
C
C
P

Chapter 8:

Traceability

Traceability is not about passing the blame to someone else.

It is about finding the source and protecting the consumer.

Traceability does not make food safe. It is a risk management tool to be used in order to assist in containing a food safety problem
(EC Guide to General Food Law Requirements).

Traceability has become a major item in the European political agenda. It applies to all in the chain of supply. The law is simple.

> **You must have records of all foods going out and coming in to your business that you can show to the authorities on demand.**

8.1 Why traceability?

Experience has shown that the functioning of the internal market in food and feed can be jeopardised where it is impossible to trace food and feed. It is therefore necessary to establish a comprehensive system of traceability within the food and feed businesses so that targeted and accurate withdrawals can be undertaken or information given to consumers or control officials, thereby avoiding unnecessary wider disruption in the event of food safety problems (178/2002 recital 28).

This is brought into law by article 18 of 178/2002 which says; *The traceability of food, feed, food-producing animals, and any other substance intended to be, or expected to be, incorporated into a food or feed shall be established at all stages of production, processing and distribution.* This is therefore all embracing and includes everyone.

The article goes on to specify that FBOs shall be able to identify the person from whom they have been supplied and secondly the business to which their products have been supplied.

There must be information on the label to facilitate this traceability.

It is necessary to ensure that a food or feed business, including an importer, can identify at least the businesses to and from which the food, feed, animal or substance that may be incorporated into a food or feed has been supplied, to ensure that on investigation, traceability can be assured at all stages. A food business operator is best placed to devise a safe system for supplying food and ensuring that it is safe (178/2002 recitals 29 and 30).

At first sight this may seem to be a significant burden on business to keep such records. In fact it is not, as the requirement is only for a one-up one-down record and sales/gifts to the final consumer are exempt.

Fig. The one-up one-down principle

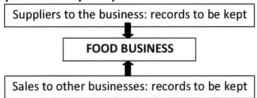

The tracing system adopted must be capable of effecting product withdrawals and recalls.

These two actions, one where the product has left your control (withdrawal) and the other, when it has reached the consumer (recall) should both be implemented quickly.

There are then three things that have to happen:

1. You must believe the product is not in compliance.
2. You must initiate the appropriate actions.
3. You must tell the competent authority.

If a food business operator considers or has reason to believe that a food which it has imported, produced, processed, manufactured or distributed is not in compliance with the food safety requirements, it shall immediately initiate procedures to withdraw the food in question from the market where the food has left the immediate control of that initial food business operator and inform the competent authorities thereof (178/2002 article 19.1). **This is withdrawal.**

This is amplified to say: *Where the product may have reached the consumer, the operator shall effectively and accurately inform the consumers of the reason for the withdrawal, and, if necessary, recall from the consumers products already supplied to them when other measures are not sufficient to achieve a high level of health protection* (178/2002 article 19.1). **This is recall** (see paragraph 8.10).

A further clause in the article says that the FBO must immediately inform the authorities if the food may be injurious to health. The last clause says that FBOs shall cooperate with the authorities.

You may inform the FSA on https://incidents.foodapps.co.uk/IncidentReportForm/login.aspx of any food safety incident.

8.2 Who is affected?

This Regulation applies to all stages of production, processing and distribution of food. It does not apply to domestic production, preparation or handling of food for private domestic consumption. It does cover any undertaking, whether for profit or not, and whether public or private, dealing with food. Food is any substance intended or reasonably expected to be ingested by humans.

It does create a new general obligation for all FBOs.

The Regulation applies to FBOs at all stages of the food chain from primary production (producing food, animals and harvesting of plants), food/feed processing to distribution. This includes charities. But the guidance does say: *that enforcers should take into consideration the particular situation of charities and donation activities in the context of enforcement and sanctions.* It does cover transporters and storage operators.

The traceability provisions of the Regulation do not have an extraterritorial effect outside the EU. Imported products must comply with EU hygiene requirements but the traceability diktat only applies from the importer. However the importer must be able to identify from whom he bought the product in the third country. It includes primary producers, manufacturers, wholesalers, retailers, transporters, distributors, caterers, food brokers and charity functions (FSA guide July 2007).

8.3 What do you have to do?

Food and feed business operators shall: be able to identify any person from whom they have been supplied with a food, a feed, a food producing animal or any substance intended to be or expected to be incorporated into a food or feed. Food and feed business operators shall: have in place systems and procedures to identify the other businesses to which their products have been supplied (article 18.2/3 of 178/2002).These systems or procedures must allow for the information to be made available to the competent authorities on demand. In Britain the Food Standard Agency has been designated as the "competent authority". Their executive arms include Trading Standards Officers and Environmental Health Practitioners (EHPs).

A further Regulation as from 27 October 2006, Para 6 under previous regulations, requires traceability for any materials that can reasonably be expected to come into contact with foods. This is explained under contact materials below in 8.11.

Traceability does not itself make food safe. It is a risk management tool to be used in order to assist in containing a food safety problem. It has the objective of enabling unsafe food or feed to be withdrawn from the market before any damage is done.

Under the previous regulations safe food or feed was determined by end product testing. By the time that had been done the product had often been distributed to retailers, other processors or even the final consumers.

The new regulations have changed that approach by insisting on quality and conformity checks to be done at each critical stage of any process. Every food business must have in place, implement and maintain a permanent procedure based on the principles of Hazard Analysis And Critical Control Points (HACCP). By this method all the relevant tests have been done before a product leaves the immediate control of the business. Thus everything that is placed on the market should be safe. But mistakes do occur and certain procedures are laid down to minimise the risks to the final consumer.

8.4 End product testing – is not a suitable instrument for guaranteeing the safety of the food and a robust HACCP system needs to be in place.

8.5 The Food Business Operator (FBO) – shall collaborate with the competent authorities on action taken to avoid or reduce risks posed by a food, which they have supplied. So, if a mistake occurs, the FBO must immediately inform the competent authority. This means both the local Environmental Health office and the Food Standards Agency. This is if he considers or has reason to believe that a food that he has placed on the market may be injurious to health. The FBO must also say what action he has taken to prevent risks to consumers. He must cooperate with the authority.

It is of significance here to realise that most "unsafe" products are either discarded within the premises or reworked to make them safe. Such incidents are not covered by the Regulation, which specifically says "placed on the market". That implies that the product has left the immediate control of the business.

8.6 What is unsafe?

Food shall be deemed to be unsafe, if it is considered to be either injurious to health or unfit for human consumption. Regard here shall be had to the normal conditions of use of the food by the consumer and at each stage of production, processing or distribution and to the information provided to the consumer by way of the label or other information generally available (Article 14 of 178/2002).

For instance, raw chicken may be slightly contaminated, but it will be cooked before eating. A similar contamination in chocolate would be unacceptable. "Injurious to health" means not only the probable short or long term effects of that food on the health of the person consuming it, but also to subsequent generations. Also it means the probable cumulative toxic effects.

"Injurious to health" can also mean the particular sensitivities of a specific category of consumers where the food is intended for that category. This will include food and drink aimed at infants, children, pregnant women, aged persons and of course those with allergies or other special needs.

This means that, if the labelling is not in conformity with the regulations, the food is deemed to be unsafe. For instance, if an allergen is included in the ingredients and not clearly stated on the label, the food must be considered unsafe (see chapter 14).

Feed shall be deemed to be unsafe for its intended use, if it is considered to have an adverse effect on human or animal health or even make the food derived from food-producing animals unsafe for human consumption.

If, at any stage of production, processing or distribution, a part of a batch is found to be unsafe, it must be assumed that the whole of that batch is also unsafe.

For the purposes of traceability, illegal foods are also considered unsafe. Thus any food containing a genetically modified ingredient that is not declared on the label is "unsafe". Similarly any food containing a banned or illegal substance, like the Sudan dyes, is unsafe.

8.7 Should you have an Internal traceability system?

The Regulation does not compel FBOs to establish a link between incoming and outgoing products. This is known as internal traceability. Nor is there any requirement for records to be kept identifying how batches are split and combined within a business to create particular products or new batches. But those who do not may run the risk of having to withdraw all their production in the event of a food safety incident rather than just the one product or batch.

So, whether or not to have a system of internal traceability is largely a commercial decision based on the possible costs of running the system against the potential costs of total withdrawal. The costs of running a system may be minimal or could be significant, if a system is employed that does not give parallel benefits. Such a system may provide stock control, order schedules, invoicing and other general management information (see Useful Links on Food Solutions website www.food-solutions.org .)

The official guidance says that FBOs should be encouraged to develop systems of internal traceability designed in relation to the nature of their activities (processing, storage, distribution etc.) The level of detail required should be the decision of the FBO "commensurate with the nature and size of the business".

8.8 What Information should be kept?

Whilst the Regulation itself does not specify what information has to be kept the official guidance is quite clear. The minimum for supplies is the name and address of the supplier, nature of the product and the date of the transaction or delivery. The minimum for deliveries is again the name and address of the business, the nature of the product and the date of transaction or delivery. It is highly recommended to also include information relating to the volume or quantity, batch numbers and more details of the product (pre-packed, bulk, raw, processed etc.). The information to be kept has to be chosen in the light of the food business activity and the characteristics of the system.

Because the Regulation says that a FBO shall immediately inform the competent authorities, if he considers that a food he has placed on the market may be injurious to health, any system of traceability must have a quick response time. It is not sufficient to rely on invoices for records of delivery, as they may not be raised at the time of delivery. Any delay in providing information may endanger the ability to control the mistake and undermine the prompt reaction in the case of crisis. Such a "crisis" could be to prevent a faulty product moving on to a third party or eventual consumer.

The official guidance says that the minimal data, as defined above, shall be available immediately to the competent authorities. The highly recommended data shall be available as soon as reasonably practicable.

8.9 How long do you need to keep records?

- For highly perishable products with a "use by" date less than three months, records should be kept for six months from the date of manufacture.
- For products without a specified shelf life or one of more than three months, records should be kept for five years.
- For products with a shelf life of more than five years, records should be kept for the period of the shelf life plus six months.

There is no specific guidance for fine wines.

8.10 Withdrawals and recalls

Withdrawing and, if necessary, recalling a product can be very expensive. If the product has gone to a retailer, but not been displayed for sale, you will not be paid for the delivery. You will pay for transport back to your premises and you may face a handling charge. If the product has been offered for sale, you may be faced with extra charges for advertising and compensation to members of the public that may have purchased units. In either event you must have notified the local Environmental Health Department and the Food Standards Agency. In both events they will have published details of the misdemeanour and your reputation may be tarnished.

In Britain the Food Standards Agency operates a rapid alert system that warns Local Authorities of any possible hazard. The Agency is also responsible for notifying other Member States and the European Food Safety Authority who may, if they consider it necessary, notify other countries. The FBO only has the responsibility to notify his Local Authority and the FSA.

The FSA system uses email for its alerts and issues updates at regular intervals so that interested parties can be kept informed of products, batch numbers or dates with pictures of suspect packaging.

If a FBO considers or has reason to believe that a food, which he has imported, produced, processed, manufactured, or distributed is not in compliance with the food safety requirements, he shall immediately initiate procedures to withdraw. He shall withdraw the food in question from the market where the food has left the immediate control of that initial FBO and inform the competent authorities. Where the product may have reached the consumer, the FBO shall accurately and effectively inform the consumers of the reason for the withdrawal and, if necessary, recall from consumers products already supplied to them (article 19.1 of 178/2002).

Any FBO who is a retailer must pass on relevant information to trace that product and cooperate with suppliers and the competent authorities. No one shall prevent or discourage any person from cooperating where this may reduce a risk arising from the food.

There are two levels of urgency here.

The first is when the product is not in compliance.

The second, and more severe, is when the product may constitute a hazard to health.

Withdrawal may take place at any point along the food chain from a sick beast to stocks in a café. Again the words "that initial operator" imply that the food has left his control and the remedy must involve third parties. Here the full process must be followed. Where there is the possibility for the FBO to remedy the non-compliance by their own means, without the need to require cooperation from other operators, the obligations of the withdrawal procedure do not apply.

As quoted above, a food may be classed as safe, even if it is unfit in its present form for human consumption. The wording is *"Regard here shall be had to the normal conditions of use of the food by the consumer and at each stage of production, processing or distribution and to the information provided to the consumer by way of the label or other information generally available"*. Thus raw or unprocessed foods should not be considered injurious to health, even if to eat them would make a person unwell.

In practice experience has shown that many product withdrawals are the result of carelessness and human error. Not all are life threatening. Some involve durability date errors. Others are from the break-up of plastics components in the production process. More serious ones have been incorrect labelling when allergens have been included in the ingredients, but not declared on the label. The infamous inclusion of Sudan dyes in a mass of products was more a question of illegality rather than safety. It was said that a person would have to drink some 270 litres of Worcester sauce to become seriously ill from the dye. He might not survive the sauce.

There have been several cases of deliberate sabotage by aggrieved employees or political activists. These are hard to detect, as the contamination may not be something that routine testing should pick up. Irregular and intermittent faults are difficult to diagnose.

The hazard analysis prior to devising the control system based on the principles of HACCP should alert the FBO to any reasonable possibilities. It cannot be stressed too much that, with proper risk analysis and control during processing, the withdrawal and recall procedures should never need to be invoked. There are some obvious anomalies. For instance:

- A small FBO, such as a bed and breakfast establishment, may buy some products from a supermarket. The supermarket may not know that the buyer represents a business and so may not have a record of that sale.
- When animals are brought in for slaughter the only identifying mark on them are the ear tags. One of the first operations is to cut off the head with the attached ears, at which point the identity may be confused.

8.11 The second traceability regulation covers packaging and contact materials

Regulation EC 1935/2004 demands that materials and articles that may come into contact with foods must be traceable at all stages. This is to facilitate control, the recall of defective products, consumer information and attribution of responsibility. The Regulation has been prompted partly by the practice of "intelligent" packaging that may alter the properties of a food during storage.

Two specific definitions apply to this Regulation:

1. Traceability is to follow a material or article through all stages of manufacture, processing and distribution.
2. "Placing on the market" here means the holding of materials and articles for the purpose of sale whether free of charge or not.

The materials covered include rubbers, ceramics, plastics, paper, glass, metals, inks, textiles, waxes, cork and wood etc.

The Regulation covers several aspects to do with contact materials such as safety and specific labelling. Article 17 defines the requirement for tracing where it is technologically feasible. *FBOs shall have in place systems and procedures to allow the identification of the businesses from which and to which materials, articles or other substances are supplied. These materials, which are placed on the market, shall be identified by an appropriate system, which allows their traceability by means of labelling or relevant documentation or information.*

There are rules for wrapping and packaging materials set out in annex II of 852/2004. *Materials used for wrapping and packaging are not to be a source of contamination. Nor are they to be stored where they could be contaminated. Further legislation is being prepared, particularly for plastics materials.*

There are also European Standards approved by the Commission for Européen Normalisation (CEN) that define the materials that come into contact with foods in plant and machinery. These are split into two, those that are in physical contact and those in splash areas. The standards even specify the alloy content of the stainless steels to be used. It is enough for the average business to insist on the CE mark being on any plant or machinery bought. In this way the FBO can be confident that the equipment not only conforms to hygiene standards, but is also safe to use.

8.12 The practical application of Traceability Regulations

This depends very much on the nature and size of the business. The Regulation rightly does not specify a system but only the required outcome.

A small café will have very different problems than a major supermarket. Both will need the same outcome, but the way they get there will vary. Those undertakings selling to another business will need records of those movements as well as their purchases. We cannot here go into details of all systems available, but give a few examples of what some have found to work.

The basic system is no more than a box into which all delivery notes are put. You will need till receipts as well to have a record of consumables. This system may work for some very small undertakings but has many shortfalls.

The recent Sudan dye recall involved purchases at least two years previously; so finding the records in a hurry could be difficult. The system could be made more sophisticated by recording everything in a daybook. This would certainly make accounting easier as stock movements could be ticked off as and when payments are made or received.

The idea of incorporating the traceability requirement into essential business procedures like accounting and stock control means that it becomes less of a chore and may almost appear automatically. It may also improve management control of the business. There are several systems available.

Some use bar codes and readers. Others use Radio Frequency Identification (RFID), which automatically tells a main computer where everything is.

8.13 Traceability summary

Of the three things that the new regulations demand of all food businesses, Registration, Traceability and Risk control, this chapter has sought to give an introduction to the second. The basic law is simple.

> **Every food business must have records, instantly available, of all goods coming in and going out to other businesses.**

The requirements that we have explained in this chapter are as set out in the regulations and official guidance. Many commercial traders may require more, especially from importers. Also the requirements of accreditation standards may demand full internal traceability as defined in ISO 22,000 or the Global standard.

You need to ask yourself the following questions:

Can you demonstrate that you can identify where all your ingredients came from at short notice? *Yes or No*	
Can you demonstrate that you can identify the businesses that you have sold product to? *Yes or No*	
Are you using a system that just gives compliance, but does not help you in your business? *Yes or No*	

Chapter 9:
Premises

If your buildings are not right, not only could your business be inefficient, but it could also generate hazards.

There are three main areas of regulation that govern your premises. The first is contained in the Food Hygiene Regulation EC 852/2004, the second is the Health and Safety raft of regulation and the third comes under Environmental rules. All premises construction also comes under planning rules, which vary from one part of the country to another depending on local policies. This chapter deals primarily with the food safety aspects.

Premises means any establishment, any place, vehicle, stall or moveable structure and any ship or aircraft (SI 2006 No 14). This **Statutory Instrument** (see glossary) called the Food Hygiene (England) Regulations 2006 has no reference to premises other than to say that you must comply with the general hygiene requirements laid down in annex II of 852/2004. So the document brings into English law the provisions of the EU Regulations and defines the penalties for transgressions. Here an establishment also means any unit of a food business.

9.1 The Premises laws

The requirements for premises were contained in the UK Food Safety Act 1995 SI 1995 no 1763. This has now been revoked and superseded by the EC Regulation 852/2004.

Further requirements are set out in Workplace (Health, Safety and Welfare) Regulations 1992, which define some of the provisions of the EC regulation. Some reference also should be made to the European Code of Practice on cleanability of commercial food equipment. This document runs to 150+ pages and is the CEN standard. This is the definitive description of materials and test procedures for all equipment and fittings within food premises.
References are made to it in the comments below.

There are also specific requirements for premises dealing with products of animal origin. These are covered in EC 853/2004. For the most part they are identical to annex II quoted below except for the needs of slaughter houses and fishery vessels. The FSA has produced guidance on this in their Guide to Food Hygiene and Other Regulations for the UK Meat Industry.

The detailed requirements are set out in annex II of 852/2004 and are qualified in article 4 which says *"Food business operators carrying out any stage of production, processing or distribution of food ...shall comply with the general hygiene requirements laid down in annex II"*.

The annex has three chapters, which are shown in 9.2, 9.3 and 9.4, which define requirements in general, in rooms where foods are prepared, treated or processed and thirdly in moveable or temporary premises.

9.2 General requirements for food premises (other than those specified in chapter 9.4 temporary premises)

1. *Food premises are to be kept clean and maintained in good repair and condition.*
2. *The layout, design, construction, siting and size of food premises are to:*
   a) *permit adequate maintenance, cleaning and/or disinfection, avoid or minimise air-borne contamination and provide adequate working space to allow for the hygienic performance of all operations;*
   b) *be such as to protect against the accumulation of dirt, contact with toxic materials, the shedding of particles into food and the formation of condensation or undesirable mould on surfaces;*
   c) *permit good food hygiene practices, including protection against contamination and, in particular, pest control;*
   d) *where necessary, provide suitable temperature-controlled handling and storage conditions of sufficient capacity for maintaining foodstuffs at appropriate temperatures and designed to allow those temperatures to be monitored and, where necessary, recorded.*

> **Comment: temperature control is a basic requirement of all food processing to ensure that bacteria, yeasts and moulds do not rapidly multiply. Temperature control requirements are set out in Schedule 4 of the 2006 Food Hygiene (England) Regulations and are dealt with in chapter 15.**

3. *An adequate number of flush lavatories are to be available and connected to an effective drainage system. Lavatories are not to open directly into rooms in which food is handled.*

> **Comment: lavatories do not need to be joined on to the food premises but may be in a separate building. If they are, it is good practice to have another changing and washing facility adjacent to the food rooms.**

4. *An adequate number of washbasins is to be available, suitably located and designated for cleaning hands. Washbasins for cleaning hands are to be provided with hot and cold running water, materials for cleaning hands and for hygienic drying. Where necessary, the facilities for washing food are to be separate from the hand-washing facility.*

5. *There is to be suitable and sufficient means of natural or mechanical ventilation. Mechanical airflow from a contaminated area to a clean area is to be avoided. Ventilation systems are to be so constructed as to enable filters and other parts requiring cleaning or replacement to be readily accessible.*

6. *Sanitary conveniences are to have adequate natural or mechanical ventilation.*

> **Comment: lights, particularly fluorescent and strip lights, must have protection to prevent any glass dropping into foods below. Also all windows and skylights in a workplace shall be of a design or so constructed that they can be cleaned safely (Health and Safety Regs).**

7. *Food premises are to have adequate natural and/or artificial lighting.*

8. *Drainage facilities are to be adequate for the purpose intended. They are to be designed and constructed to avoid the risk of contamination. Where drainage channels are fully or partially open, they are to be so designed as to ensure that waste does not flow from a contaminated area towards or into a clean area, in particular an area where foods likely to present a high risk to the final consumer are handled.*
9. *Where necessary, adequate changing facilities for personnel are to be provided.*
10. *Cleaning agents and disinfectants are not to be stored in areas where food is handled.*

9.3 Specific requirements in rooms where foodstuffs are prepared, treated or processed *(excluding dining areas and those premises specified as temporary or movable in chapter 9.4):*
1. *In rooms where food is prepared, treated or processed (excluding dining areas and those premises specified in Chapter 9.4, but including rooms contained in means of transport) the design and layout are to permit good food hygiene practices, including protection against contamination between and during operations. In particular:*
   a) *floor surfaces are to be maintained in a sound condition and be easy to clean and, where necessary, to disinfect. This will require the use of impervious, non-absorbent, washable and non-toxic materials, unless food business operators can satisfy the competent authority that other materials used are appropriate. Where appropriate, floors are to allow adequate surface drainage;*
   b) *wall surfaces are to be maintained in a sound condition and be easy to clean and, where necessary, to disinfect. This will require the use of impervious, non-absorbent, washable and non-toxic materials and require a smooth surface up to a height appropriate for the operations unless food business operators can satisfy the competent authority that other materials used are appropriate.*

Comment: Wall coverings should be of a light colour to permit the easy identification of dirt or contamination.

---

c) ceilings or, where there are no ceilings, the interior surface of the roof and overhead fixtures are to be constructed and finished so as to prevent the accumulation of dirt and to reduce condensation, the growth of undesirable mould and the shedding of particles;

d) windows and other openings are to be constructed to prevent the accumulation of dirt. Those, which can be opened to the outside environment, are, where necessary, to be fitted with insect-proof screens, which can be easily removed for cleaning. Where open windows would result in contamination, windows are to remain closed and fixed during production;

e) doors are to be easy to clean and, where necessary, to disinfect. This will require the use of smooth and non-absorbent surfaces unless food business operators can satisfy the competent authority that other materials used are appropriate;

> **It is good practice to fit bristle strips under doors to prevent rodents getting in.**

f) surfaces (including surfaces of equipment) in areas where foods are handled and in particular those in contact with food are to be maintained in a sound condition and be easy to clean and, where necessary, to disinfect. This will require the use of smooth, washable corrosion-resistant and non-toxic materials, unless food business operators can satisfy the competent authority that other materials used are appropriate.

> **Comment: separate recommendations are for food contact surfaces and splash surfaces. There are EN Standards for all surfaces taking into account pH, temperature, cleaning procedures, pressure and thermal stress, condensation, radiation and mechanical attrition or erosion. See CEN/TR 15623.**

2. Adequate facilities are to be provided, where necessary, for the cleaning, disinfecting and storage of working utensils and equipment. These facilities are to be constructed of corrosion-resistant materials, be easy to clean and have an adequate supply of hot and cold water.

3.   Adequate provision is to be made, where necessary, for washing food. Every sink or other such facility provided for the washing of food is to have an adequate supply of hot and/or cold potable water (see glossary) consistent with the requirements of Chapter 10.2 and be kept clean and, where necessary, disinfected.

9.4 Requirements for movable and/or temporary premises (such as marquees, market stalls, mobile sales vehicles), premises used primarily as a private dwelling-house, but where foods are regularly prepared for placing on the market and vending machines:

1.   Premises and vending machines are, so far as is reasonably practicable, to be so sited, designed, constructed and kept clean and maintained in good repair and condition as to avoid the risk of contamination, in particular by animals and pests.

2.   In particular, where necessary:

   a)   appropriate facilities are to be available to maintain adequate personal hygiene (including facilities for the hygienic washing and drying of hands, hygienic sanitary arrangements and changing facilities);

   b)   surfaces in contact with food are to be in a sound condition and be easy to clean and, where necessary, to disinfect. This will require the use of smooth, washable, corrosion-resistant and non-toxic materials, unless food business operators can satisfy the competent authority that other materials used are appropriate:

   c)   adequate provision is to be made for the cleaning and, where necessary, disinfecting of working utensils and equipment;

   d)   where foodstuffs are cleaned as part of the food business' operations, adequate provision is to be made for this to be undertaken hygienically;

   e)   an adequate supply of hot and/or cold potable water is to be available;

   f)   adequate arrangements and/or facilities for the hygienic storage and disposal of hazardous and/or inedible substances and waste (whether liquid or solid) are to be available;

g) adequate facilities and/or arrangements for maintaining and monitoring suitable food temperature conditions are to be available;

h) foodstuffs are to be so placed as to avoid the risk of contamination so far as is reasonably practicable.

The terms "adequate", "where necessary" and "sufficient" shall mean just that to achieve the objectives of the regulations (852/2004 article 2.3). The objective is to provide a high level of health protection. Such health protection is aimed at the consumers of the foods. The health protection of you and your staff and anyone who comes on to your premises as a customer or visitor is the concern of the Health and Safety regulations.

Typical Cases where premises were not to the required standard

- **Floor and wall coverings that were not washable or were painted with inappropriate paints.**
- **Domestic cleaning materials being used in a commercial premises, which were not suitable for the sorts of dirt found in the building.**
- **Nearly all the prosecutions from one authority were for dirty premises.**
- **Very rare to find a changing area where staff could leave their non-work clothes.**

It is well worth getting someone, usually a professional, to look round your premises at least once a year to pin-point any areas of hygiene risk that you may have overlooked. Experience has shown that areas, which have been improved a bit, will look good to the owner, but may still be below normal standards and even pose significant risk. The owner will only see such areas as improved. They may not be good enough.

9.5 Fire risks

Since October 2006 all non-domestic premises must comply with the Fire Safety Order. This abolishes the old fire certificates and demands that the "responsible person", usually the business owner, must have done and put in writing a fire risk assessment. Your fire risk assessment will help you to identify the risks that can be removed or reduced and to decide the nature and extent of the general fire precautions you need to take to protect people against the fire risks that remain. If you employ more than five people, you must record the significant findings of the assessment. For more complex premises you should use a competent person to do the assessment, usually a consultant or other specialist with the appropriate experience and knowledge. For advice contact the Institution of Fire Engineers on www.ife.org.uk.

The assessment will look at the type of building, the likely occupants and visitors, especially those with disabilities, the fire hazards and measures for their control like ignition sources, means of escape, warning systems, emergency lighting, extinguishing appliances, smoke detectors, staff training and practices.

Together this will form the basis of an action plan.

Do not forget your commercial records and customer/supplier details as these could be critical to your business in the event of even a very small fire. The use of third party certificated firms is generally held to constitute material evidence of due diligence on your part.

When did you last look at all your premises critically? *Date*	
Is your pest control and fire risk assessment up to date? *Yes or No*	

Chapter 10: Equipment

"Give us the tools and we will finish the job"
- Churchill.
This chapter looks at the things that go into your buildings.
The rules are not what some people think; there is flexibility.

Almost all equipment used in commercial premises – and in most domestic homes – is now made to exacting standards set by national and international bodies. Very simply there are three levels.

- The ISO standards are international.
- The CEN (the Commission Européen Normalisation) set standards for almost everything. You will have seen the letters CE on electrical goods.
- The BSI (the British Standards Institute) has one seat on the CEN and at the last count had over 1300 standards to do with food.

The value of these standards cannot be over-rated. Wherever you send a sample to be tested for say Listeria or heavy metal contamination you can be confident that the tests will be done to international standards and methods. Most standards we take for granted now, but just think what it was like when each railway company had its own gauge rails or if each nation had incompatible telephone systems.

The process of developing standards (elaborating is the European word) is essentially simple. A committee looks at say a class of equipment, assesses the risks involved from data and experience and writes a draft standard for specifications, which is circulated to anyone interested. Everyone, including you, has the right to comment and make amendments, but usually it is the experts in design, manufacture and use that are most vociferous. The amended draft is voted on by the experts with each state having one vote. The BSI votes for Britain.

Standards for equipment cover three major areas.

The first is for operator safety. You are well advised to ensure that all your electrical equipment has the CE mark. The standards even cover such things as the size of the hole on mincing machines to prevent people sticking their fingers in.

The second is fitness for purpose. The standards define claims that can be made according to the use to which the equipment may be put. Thus a microwave oven designed for occasional domestic use will comply with one standard for performance, which may be totally inadequate for commercial use.

The third area, particularly relevant in the food industry, is hygiene. There are many different stainless steels but only a few of them are suitable and safe for food contact. There are many sorts of cling film, but only some are safe for food.

The advice is simple. When buying, hiring or using equipment, ensure that it has been made and set up to the right standards. Look for the CE mark.

Flexibility is also appropriate to enable the continued use of traditional methods at any of the stages of production, processing or distribution of food and in relation to structural requirements for establishments (recital 16 of 852/2004).

It goes on to emphasise that flexibility should not compromise food hygiene objectives. **This means that in certain circumstances wood may be used rather than stainless steel or plastic.**

It is recommended that chopping boards should be colour coded for different uses - red for fresh meat, white for ready to eat foods, blue for fish and green for vegetables. This is not legislation and so not mandatory. The important thing is to keep them clean and replace them if they become damaged.

The law on equipment is quite simple and is set out in chapter V of 852/2004.

10.1 Equipment requirements

1. *All articles, fittings and equipment, with which food comes into contact, are to:*
   a) *be effectively cleaned and, where necessary, disinfected. Cleaning and disinfection are to take place at a frequency sufficient to avoid any risk of contamination;*
   b) *be so constructed, be of such materials and be kept in such good order, repair and condition as to minimise any risk of contamination;*
   c) *with the exception of non-returnable containers and packaging, be so constructed, be of such materials and be kept in such good order, repair and condition as to enable them to be kept clean and, where necessary, to be disinfected;*
   d) *be installed in such a manner as to allow adequate cleaning of the equipment and the surrounding area.*
2. *Where necessary, equipment is to be fitted with any appropriate control device to guarantee fulfillment of this Regulation's objectives.*
3. *Where chemical additives have to be used to prevent corrosion of equipment and containers, they are to be used in accordance with good practice.*

Effective cleaning is dependent on at least six criteria:
1. Temperature of cleaning solutions.
2. Concentration of sterilants and other chemicals.
3. Abrasion effects, actually wiping or rubbing.
4. Time of contact with chemicals.
5. Cross contamination by dirty cloths etc.
6. Maintenance of sterile conditions, air pollution, dust etc. after cleaning.

10.2 Water

There are also strict rules about the use of water. It must be of drinkable quality. The word used is potable. Any recycled water that has been used for cooling or ice making should not come into contact with foods.

10.3 Microwaves

The output from some domestic microwave ovens may drop by 25% on the second use and 50% on the third use within an hour. This could mean that your product is not reheated or cooked as much as you thought it had been.

It poses a risk. As a general rule a microwave only oven is for reheating while a combination microwave oven is for reheating and primary cooking.

A commercial oven is built to withstand hard use every day.

There are four main power groups:

- Light duty ovens for occasional use, often in domestic homes.
- Medium duty ovens are for small restaurants and cafés and are usually 1100 to 1500 watts.
- Heavy duty are for busy pubs and hotels.
- Extra heavy duty ovens are for major catering use where large quantities of food need cooking at once.

Microwave ovens need regular professional service as well as daily cleaning. Cleaning prevents the inside of your oven becoming covered in baked food which is another hazard. Be careful to use only soft cloths and approved cleaners so you do not damage the surface.

Typical Cases where equipment was not used correctly or not to standard
<ul><li>Hand washing basins used for supporting the radio, so unusable.</li><li>Hand washing basins used for washing items of clothing.</li><li>Fridge door seals caked with grease, crumbs and mould.</li><li>Where colour coding is used, it is rarely adhered to in practice.</li><li>One business had a cleaning schedule that he put up on a white board every week. This worked fine until he had to demonstrate the cleaning schedule. He could not as it was wiped off on completion.</li></ul>

Are you cleaning your equipment properly? *Yes or No*	
Is there any risk of cross contamination? *Yes or No*	

Chapter 11: Personal hygiene

Personal hygiene is important and is covered in the legislation in chapter VIII of annex II of 852/2004.

Remember that:

> *Primary responsibility for food safety rests with the food business operator.*
> **(852/2004 chapter 1, article 1.1.a)**

Under the food regulations the only admissible defence is one of **due diligence** (see chapter 1.3).

You must be able to demonstrate effectively that you **have done** what **you say you have done**. This means having in place procedures for this crucial issue.

The regulations say:

1. *Every person working in a food-handling area is to maintain a high degree of personal cleanliness and is to wear suitable, clean and, where necessary, protective clothing.*

2. *No person suffering from, or being a carrier of a disease likely to be transmitted through food or afflicted, for example, with infected wounds, skin infections, sores or diarrhoea is to be permitted to handle food or enter any food-handling area in any capacity, if there is any likelihood of direct or indirect contamination. Any person so affected and employed in a food business and who is likely to come into contact with food is to report immediately the illness or symptoms and, if possible, their causes to the food business operator.*

3. *It is good practice for staff to contain their hair with suitable coverings and not wear jewellery (Safer Food Better Business).*

If you want help with how to put in place procedures we suggest you contact your local Environmental Health department, Business Link, the SALSA team, your trade organisation guide to good practice or one the independent food advisors.

How do you check that all hands are washed regularly?

Chapter 12:
Health and Safety

"If you only think you are safe, you are likely to fall".
A brief look at the rules that protect you, your staff and your visitors.

This is a major topic and we can only give an outline summary here. The main provisions are in the Health and Safety at Work Act 1974 as amended and the following extracts are taken from that document. The full text can be seen at www.hse.gov.uk , which is the site of the Health and Safety Executive. This government body has the responsibility of enforcing the legislation.

There are three main duties. They are:

- of employers to their employees;
- of employers for visitors;
- of employees to themselves and colleagues.

12.1 General duties of employers to their employees

1. *It shall be the duty of every employer to ensure, so far as is reasonably practicable, the health, safety and welfare at work of all his employees.*

2. *Without prejudice to the generality of an employer's duty under the preceding subsection, the matters to which that duty extends include in particular;*

   a) *the provision and maintenance of plant and systems of work that are, so far as is reasonably practicable, safe and without risks to health;*

   b) *arrangements for ensuring, so far as is reasonably practicable, safety and absence of risks to health in connection with the use, handling, storage and transport of articles and substances;*

   c) *the provision of such information, instruction, training and supervision as is necessary to ensure, so far as is reasonably practicable, the health and safety at work of his employees;*

d) so far as is reasonably practicable as regards any place of work under the employer's control, the maintenance of it in a condition that is safe and without risks to health and the provision and maintenance of means of access to and egress from it that are safe and without such risks;

e) the provision and maintenance of a working environment for his employees that is, so far as is reasonably practicable, safe, without risks to health, and adequate as regards facilities and arrangements for their welfare at work.

Except in such cases as may be prescribed, it shall be the duty of every employer to prepare and, as often as may be appropriate, revise a written statement of his general policy with respect to the health and safety at work of his employees and the organisation and arrangements for the time being in force for carrying out that policy, and to bring the statement and any revision of it to the notice of all his employees.

12.2 General duties of employers and self-employed to persons other than their employees

It shall be the duty of every employer to conduct his undertaking in such a way as to ensure, so far as is reasonably practicable, that persons not in his employment who may be affected thereby are not thereby exposed to risks to their health or safety.

12.3 Duties of Employees

It shall be the duty of every employee while at work to take reasonable care for the health and safety of himself and of other persons, who may be affected by his acts or omissions at work.

The text of the Act runs to 129 pages and covers all sectors of industry. It sets up inspectors who have powers of entry and can impose improvement notices or even prohibition notices. It is a criminal offence to contravene the regulations.

12.3 Practice

In practice you should nominate someone, could be yourself, to take responsibility. This should be recorded as evidence.

- If you have more than five employees, you must have a written policy statement.

- You must also be insured for at least £5 million and display the certificate.
- You should also have some posters on display as well as the no smoking signs.
- You will need to have emergency plans well developed and recorded. This will include evacuation of the premises, exits etc.
- You have to have a first aid kit readily available and checked regularly. One person at least should have some training and be designated.
- You must report any accident at work if it results in three days off or any dangerous occurrences – where something happens that does not result in an injury, but could have done (near miss). Reporting is done on 0845 300 99 23.
- You must also have a record of every injury. Full details of procedures can be found at www.riddor.gov.uk. Reporting is done on 0845 300 99 23. RIDDOR is the Reporting of Injuries, Diseases and Dangerous Occurrences Regulation 1995.
- You must have done risk assessments for all aspects of your business. See chapter 3 on risk and 7 on HACCP. You do not need a fire certificate now, but you must comply with the Regulatory Reform (Fire Safety) Order 2005. This too involves a risk assessment and, if you employ five or more people, you must have it written down. For all the details and official guidance go to www.communities.gov.uk and click on Fire (see chapter 9.5).
- You must comply with the Electricity at Work Regulations: see www.hse.gov.uk for details and which equipment has to be regularly tested.
- You should have all your gas equipment checked every year. Many insurers will not cover you unless you do.
- Your risk assessment must take in premises (see chapter 9). Whilst that chapter is specific to food premises for hygiene hazards, you must also look at any hazards to yourself, your staff and your visitors.

Two other regulations apply here.

The first is COSHH - the Control of Substances Hazardous to Health Regulations 2002. This applies particularly to cleaning chemicals, which may present risks to staff and customers/consumers, if there were any contamination.

The second is REACH - the Recording, Evaluation and Authorisation of Chemicals. This is more for those who use or store quantities over one tonne per year.

There are some obligatory signs in blue that you must display like "Wash your hands" and "Wear goggles" etc.

There are a mass of regulations that you could fall foul of in certain circumstances. Only you know your business and can do the risk assessment to protect persons from hazard. Asbestos, noise, heat, radiation, chemicals, zoonoses (see glossary), heavy metals, environmental and many other factors all have specific regulations which you may need to know about. The Health and Safety Executive local office should be able to help or see www.hse.gov.uk.

Facts

- **There have been more injuries reported in the food industry than in the construction sector over the last few years.**
- **This is becoming worse with a 9% increase last year, although major injuries have reduced by a third since 1990.**
- **If one of your staff, unseen by you, is spotted using a ladder wrongly, you could be fined even though there had been no accident.**
- **Slips and trips are said to be the most common cause of injury.**
  It could cost you dearly.

Reviewing Risk

What are the risks to your staff?
What are the risks to you visitors/customers?
What are the risks to you visitors/customers?
What are you doing to minimise those risks?

Chapter 13:
Labelling

An introduction to the current rules on food labelling.

The laws on food labelling are under review. The basic concept is contained in article 16 of 178/2002, which says: *Without prejudice to more specific provisions of food law, the labelling, advertising and presentation of food or feed, including their shape, appearance or packaging, the packaging materials used, the manner in which they are arranged and the setting in which they are displayed, and the information which is made available about them through whatever medium, shall not mislead the consumer.* The following table illustrates some of the details for pre-packaged foods.

L
A
B
E
L
L
I
N
G

Mandatory	Voluntary
Name	Vegetarian/vegan
List of Ingredients	Assurance Scheme
Instructions for use (if failure to include might mislead)	Nutrition Information (if no claims are made)
Net Quantity (weights and measures)	May contain (traces of nuts etc)
Date of minimum durability; use by or best before dates	Special offers, competitions, BOGOF etc
Name and address of manufacturer, packer, seller	Method of slaughter (Halal, Shechita)
Nutrition panel (where a nutrition claim is made)	Animal welfare conditions (free-range)
Raw milk labelling	Nutrition signposting
Allergen information (in the ingredient list)	Brand information
Alcoholic strength by volume (drinks over 1.2% only)	Marketing claims, such as "no artificial additives" and similar
Quinine labelling	Environmental impact (dolphin-friendly)

Mandatory	Voluntary
High caffeine content warning (drinks containing over 15mg/l caffeine	Production methods (organic etc)
Sweetener labelling ("with sweeteners")	Guideline daily amounts
Country of origin (only if not to do so might mislead)	Country of origin (where not required)
PKU warning ("contains a source of phenylanine")	Logos (red tractor etc)
Packaging gases ("packed in a protective atmosphere")	Marketing terms, fresh, pure, natural
Polyol warning ("excessive consumption may produce laxative effects")	Customary or descriptive names
GMO labelling	Quality type (e.g. 100% chicken breast)
Irradiated food labelling ("irradiate" or "treated with ionising radiation")	Pictures and graphics, including flags and icons that do not imply claims
Quantity of certain ingredients **(QUID)** (see glossary) e.g. Chicken 10%	Number of servings

Separate rules apply to foods sold loose (non-pre-packed) and to very small quantities. Labelling of allergens is covered in chapter 14. Any claims on labels must be justified by accepted scientific evidence.

This means the food must be what it says it is as well as what it looks like.

The laws covering the labelling of foods are complex and are defined in over 40 separate EC regulations as well as UK rules. The best advice is to have your label draft approved by your local Trading Standards office. If they have approved it, you are then covered in all areas of the country under the Home Authority principle (see glossary).

13.1 Claims

A claim: *means any message or representation, including pictorial, graphic or symbolic representation, in any form, which states, suggests or implies that a food has particular characteristics* (EC/1924/2006). The regulation prohibits any claim as defined above, which suggests there is a beneficial relationship between that food and health. It includes suggestions that health may be affected by not consuming a food and any reference to the rate or amount of weight loss. There will be a positive list of generic claims that can be made issued in the next few years. In the meantime certain definitions are set out in the annex to the regulation summarised as follows. A claim may not be specific as the regulation says it "could have the same meaning for the consumer" for each one of these items.

Nutrition claims and conditions applying to them:

Low energy
Can only be made when the product has less than 40kcal per 100 g for solids or 20 kcal per 100ml for liquids.

Energy reduced
Can only be made when the energy value is reduced by at least 30%.

Energy free
Can only be made if the energy value is less than 4 kcal per 100g.

Low fat
Can only be made when the product has less than 3g of fat per 100 g for solids or 1.5g per 100 ml for liquids. There is an exception for milk where the limit is 1.7g per 100 ml.

Fat free
Can only be made if the product has less than 0.5g fat per 100 g or 100 ml. Note that claims expressed as x% fat free are prohibited.

Low saturated fat
Can only be made if the sum of saturated fatty acids and trans fatty acids in the product does not exceed 1.5g per 100g or 0.75g per 100 ml for liquids.

Saturated fat free
The limits go down to 0.1g per 100g or 100ml.

Low sugars
Can only be made if the product has less than 5g of sugars per 100g or 2.5g per 100ml.

Sugar free
The limit goes down to 0.1g per 100g or 100ml.

With no added sugars
Can only be made if the product has no added mono- or di-saccharides or any other food used for its sweetening properties. If sugars are naturally present in the food the label should say "Contains naturally occurring sugars".

Low sodium/salt
Can only be made if the product contains less than 0.12g sodium per 100g or 100ml. Note that salt is 40% sodium, so 0.12g sodium is 0.3g salt.

Very low sodium/salt
Can only be made where there is less than 0.04g sodium or 0.1g salt per 100g or 100ml.

Sodium or salt free
Less than 0.005g of sodium or 0.013g salt per 100g or 100ml.

Source of fibre
Can only be made where the product contains at least 3g fibre per 100g.

High fibre
Can only be made where the product contains at least 6g fibre per 100g or at least 3g fibre per 100 kcal.

Source of protein
Can only be made where at least 12% of the energy value is provided by protein.

High protein
Here at least 20% of the energy value is provided by protein.

Source of [xxx] vitamin or [xxx] mineral
Can only be made where the product contains at least a significant amount of the material. These amounts are defined in the annex to Directive 90/496/EEC and amended by article 6 of EC/1925/2006 on

the additions of vitamins and minerals to foods.

High vitamin [xxx] or mineral [xxx]
Must have twice the value as "source of [xxx]".

Contains nutrient [xxx]
A "nutrient" means any protein, carbohydrate, fat, fibre, sodium, vitamins and minerals listed in the annex to 90/496/EEC or components thereof. A claim can only be made if the product complies with EC/1924/2006, the Claims Regulation, which limits claims to where the presence or absence of the nutrient has a beneficial effect as established by scientific evidence.

Increased nutrient [xxx]
Can only be used if the named nutrient has been increased by 30%.

Reduced nutrient [xxx]
Can only be made if the named nutrient has been reduced by 30% or 10% for some micro-nutrients or 25% for salt.

Light or Lite
Means the same as "reduced". The claim shall also have an indication of the characteristics, which make the food light or lite.

The full text of these definitions was published in the Official Journal of the EU on 18 January 2007. All decisions, directives and regulations are published in the Official Journal once they are agreed. The Regulation is EC/1924/2006. If any of these claims is made there must also be a panel showing the nutrients in the food and their quantities. The full text of the Regulation is on the Food Solutions website.

13.2 Dates on Labels
It is a legal requirement that a durability date should be put on all pre-packed foods.

> The one critical date that must be put on pre-packed perishable foods is the **USE BY** date.
> It is an **offence to sell** goods **after the USE BY date** has passed.
> To eat them could put the consumer's health at risk.

After the Use By date the food could become harmful to anyone eating it, usually through microbial deterioration. Following the date there should be storage instructions, which are usually about temperature and conditions.

The Best Before date is more about quality than safety. It indicates the date after which the flavour or texture may be impaired. Shops are allowed to sell goods that are passed their best before date. But those goods may no longer be at their best and could give rise to justified complaints.

The third date is used by retailers to help them with their stock rotation. This is a Sell By or Display Until date. It is not an offence to display goods for sale after the sell by date has passed. However, they must also have an indication of either a use by or best before date.

All date marks must be in the format and size as specified in the regulations. These details can be found on the FSA site www.food.gov.uk

Most foods that are sold loose, like cakes and buns, do not need a durability date.

Separate rules apply to very small packets. If the largest surface area of a sachet is less than 10 square cm, then, only the name of the food, any allergens and a durability date need to be put on.
Separate rules also apply to eggs. The sell by date must be no more than 21 days from the date of laying and the use by date no more than 28 days from lay.

Deciding the durability date for your product may need professional help. You must allow for temperature abuse by the customer and finish with the product still within microbiological limits. Your independent tests and comparisons should be recorded in order to demonstrate that you have taken reasonable steps to define the date. If there is the possibility that the product could become unfit, a use by date should be quoted.

LABELLING

13.3 Eggs

There are special rules for labelling eggs. The egg itself must be stamped unless you are the producer selling direct to the consumer at your own premises. In every other case any egg sold for consumption must bear a stamp that shows the method of farming, country of origin and the producer's code.

The first digit defines the method of production.
- 0 is for organic,
- 1 is for free range,
- 2 is for barn and
- 3 is for caged birds.

The code that follows is not an "8 digit code" as its length and style varies between countries (even within the UK), but must follow the overall format of a) b) c) where:
a) Is the production method code shown as 0, 1, 2 or 3.
b) Is the country of origin such as UK for United Kingdom, ES for Spain, PT for Portugal, DE for Germany and NL for Holland.
c) Is the unique producer identifier in whatever format the EU Member State desires.

In the UK the unique producer identifier is in a five digit format from England and Wales such as 12345, from Scotland it usually shows as a three digit format plus the letters SCO such as 123SCO and from Northern Ireland it normally shows as a 9 followed by a dash followed by three digits, such as 9-123.

Thus a full code for an organic farm in:
- England or Wales may appear as 0UK12345
- Scotland as 0UK123SCO
- Northern Ireland as 0UK9-123.

There is no legal obligation to put any date on an egg, but there are complex rules for the dates on the box. If you remember that the use by date must be no more than 28 days from date of lay and put no other date on the box, you should be safe.

At a recent count there are some 87 pieces of UK and EC legislation – mainly environmental – that an egg producer should be aware of.

Typical Cases of the misunderstanding of labelling rules

- I didn't really know how to establish the shelf life, so I looked at other products and had a guess.
- Milk carton design rejected because pint had a capital P.
- Most recent withdrawals and recalls have been because of a mistake in labelling.

A common mistake is to ignore stock rotation. This means that an outdated product can be accidentally used or sold. In retail establishments it has to be scrapped.

Are your labels approved by your Trading Standards Officer? *Yes or No*	
When did you last check that your durability dates were reasonable? *Date*	

Chapter 14:
Allergens

Many people have allergies that are serious or even fatal to them. You have a duty to declare certain allergens and to know enough about all of them to warn customers.

A
L
L
E
R
G
E
N
S

14.1 Legal basis

Food shall not be placed on the market, if it is considered to be unsafe. Unsafe means either injurious to health or unfit. Some foods can harm the health of some people. This is called "allergenic reaction" and for them that food is injurious to health (178/2002/EC art 14).

Under the Labelling Directive 2000/13/EC and amended by 2003/89/EC there is a legal duty to declare clearly on the label any allergen included in the ingredients.

This applies only to pre-packed foods.

Pre-packed means put into packaging before being offered for sale in such a way that the food, whether wholly or only partly enclosed, cannot be altered without opening or changing the packaging and is ready for sale to the final consumer or to a caterer, including a food which is wholly enclosed in packaging before being offered for sale and which is intended to be cooked without opening the packaging and which is ready for sale to the final consumer or caterer, but does not include individually wrapped sweets.

This is the FSA definition.

14.2 Label requirement, your duty.

Some people are so sensitive to allergens that minute quantities can trigger reactions. It is clear what to do if an allergen is deliberately included. Their presence must be declared in the list of ingredients. Its presence may also be indicated separately by means of an allergy information panel or box. It is best practice for such information to be clearly associated with the ingredients list. If such devices are employed, all allergenic foods or ingredients used in the food should be listed in the box or panel.

14.3 Accidental inclusion

In some cases traces of allergens can get into products, through inadvertent cross-contamination. The following guidance is to clarify what you should put on labels if you think there is a real possibility.

Advisory labelling should only be used when, following a thorough risk assessment, you think there is a demonstrable and significant risk of allergen cross-contamination.

It only applies to pre-packed foods. For all other foods you should be able to verbally advise customers if they ask.

Ideally you should avoid such cross-contamination (see chapter 14.5). However, in some circumstances this is impossible. If you think there is a real possibility of an allergen inclusion, it is only fair to the consumer to state that the product may include it. Then you should say "may contain xxx".

14.4 What are allergens?

Since November 2005, all pre-packed food sold in the UK has to clearly show on the label if it contains one of the following as an ingredient (or if one of its ingredients contains, or is made from, one of these):

- peanuts (these are legumes, not nuts).
- nuts (almonds, hazelnuts, walnuts, Brazil nuts, cashews, pecans, pistachios, macadamia nuts and Queensland nuts)
- eggs
- milk
- crustaceans (including prawns, crabs and lobsters)
- fish
- sesame seeds
- cereals containing gluten (including wheat, rye, barley and oats)
- soya
- celery
- mustard
- lupin
- molluscs (mussels etc.)
- sulphur dioxide/sulphites (preservatives used in some foods and drinks) at levels above 10mg per kg or per litre

Some ingredients made from these foods are no longer allergenic (e.g. refined soya oil) and so do not have to be labelled. Other foods may be added to this list in the future.

14.5 Ways of avoiding cross-contamination.

There are a number of different ways that allergen cross-contamination can occur and, with careful management, some of these risks can be avoided or reduced. All staff involved in handling ingredients, equipment, utensils, packaging and products should be aware of the situations, in which foods can be cross-contaminated by the major food allergens. Some of the areas to think about are described here:

1. The ideal approach to avoiding cross-contamination with allergens is to dedicate production facilities to specific allergenic products. Food manufacturing premises and product ranges vary greatly and this is often not an option.
2. Where possible, allergenic raw materials should be stored away from other ingredients. One way of doing this would be in sealed plastic bins that are clearly marked.
3. You should find out about the allergens in the ingredients you use and of those used by your suppliers. Any change in supplier should be accompanied by the appropriate checks.
4. Where these things are not practicable, it may be possible to separate the production of allergen-containing products from those that do not contain the allergen. This could be by making the food containing the allergen on a different day or be made at the end of the day followed by a thorough clean down.

14.6 Cleaning

Very small amounts of some allergens, such as nuts, can cause adverse reactions, which may potentially be fatal. Therefore, thorough cleaning that is effective in reducing the risks of allergen cross-contamination should be used. However, cleaning practices that are satisfactory for hygiene purposes may not be adequate for removing some allergens. There may be times when equipment may need to be dismantled and manually cleaned to ensure hard to clean areas are free from allergen residues. Even dust contamination of some allergens can cause a reaction.

Investment in developing and following appropriate cleaning regimes will help to minimise allergen cross-contamination and can reduce the likelihood of needing costly product recalls.

14.7 Packaging

Incorrect packaging and/or labelling are often the cause of allergen related product recalls. Procedures must ensure that the correct labels are applied to products and outer cases. This should be checked regularly, so that accurate information is provided to allergic consumers.

Packaging should be removed and/or destroyed at the end of a run, including any packaging that may be within the wrapping machine. This is to avoid packaging mix-ups when the product is changed.

If a product is reformulated and a new allergenic ingredient is introduced, this may lead to contamination of other products produced in the same premises, for which advisory labelling might then become appropriate.

It is only after you have taken such steps as these and still consider that there is a risk of allergen inclusion that you should print a warning.

14.8 Allergen Free foods

A growing number of food manufacturers and retailers are providing special ranges of foods made without certain common allergenic foods, such as milk, egg or cereals containing gluten. In addition, some manufacturers choose to exclude certain allergens from their premises. Even, if an advisory warning is not appropriate, this may not justify making a "Free From" or "made in allergen X free factory" claim. Consumers are likely to actively seek such products if they need to avoid particular ingredients and it is essential that any such claims are based on specific, rigorous controls. This includes checking that all ingredients and anything else that comes into contact with the food (such as packaging materials), are "free from" the particular allergen. It is also recommended that the product is tested for allergen traces by one of the specialist laboratories.

If a business produces lists of foods free from particular allergens these should regularly be reviewed and updated.

14.9 Non Pre-packed foods

Probably most serious allergic reactions occur after a person has eaten something in a catering establishment. Chef's special may be different every day and may not have a defined ingredients list. Many foods are sold loose at retail, like bread, vegetables and meats. These cannot always be labelled for very practical reasons. The European Commission and the FSA are looking at this problem and legislation is expected.

In the meantime best practice recommends that, wherever possible, allergens should be indicated on menus. Where that is impractical a member of staff should be designated to answer customer queries on allergen content. Some bread contains soya flour as well as wheat. The advice, until official guidance and/or legislation is available, is to be aware of the allergens listed above and be prepared to answer any customer with informed advice. Otherwise be frank with your customers and say that you could not guarantee there is no allergen in the product.

Practical Example

A school had a pupil who was highly allergic to peanuts. The allergic response of this child would be life threatening. A risk assessment was carried out and it was decided that there were two problems that needed addressing.

1. The chances of cross contamination could not be completely removed from the kitchen and prep areas. The solution arrived at was to stop using any nut ingredients within the dining room.
2. Some children brought packed lunches with the risk of nuts being part of the lunch pack i.e. Peanut butter etc. The solution arrived at was to ban any nut products being included in packed lunches.

This needed to be ongoing so a letter explaining the problem is included in the intro pack given to parents of new pupils.

Because the reason for the ban was explained it was accepted by all parents and pupils.

Practical Example

A firm that produced peanut free chocolates found to their dismay that their product tested positive for peanuts. After careful tracking back they found that the farmers in Sierra Leone were recycling sacks that had been used for peanuts for their cocoa beans. That was enough to produce a positive test result.

Practical Example

A girl goes to a restaurant with friends. The set menu contains shellfish. She asked for Parma ham to be substituted as she was allergic to shellfish.

The meal arrives with shell fish, the girl complains and the meal is returned to the kitchen. The chef, who was working under pressure, changes the shell fish for Parma ham on the same plate. The girl has a severe allergic reaction and dies.

Lesson to be learnt

You and your staff must be aware of the dangers and possible consequences of cross contamination.

Does every one of your staff know what allergens are? *Yes or No*	
Could someone be affected by any of your products? *Yes or No*	

Chapter 15:
Temperature control requirements

An introduction to the rules on storage temperatures.
Good Practice demands much more detail.

T E M P E R A T U R E

Many bacteria, yeasts and moulds can breed very rapidly in the right conditions. If you leave foods in those conditions, hazards can develop rather fast. Heating, as discovered by Pasteur, can kill most bacteria. However, it does not destroy the toxins (poisons) that might have been left in the food. Cooling does not kill them but only slows down their growth.

The Regulations specify that a cold chain should be maintained. This means that chilled product is not allowed to warm up and then be re-cooled or frozen.

This is specified in 852/2004: *Raw materials, ingredients, intermediate products and finished products likely to support the reproduction of pathogenic micro-organisms or the formation of toxins are not to be kept at temperatures that might result in a risk to health.*

The cold chain is not to be interrupted. However, limited periods outside temperature control are permitted, to accommodate the practicalities of handling during preparation, transport, storage, display and service of food, provided that it does not result in a risk to health - and quoted in the new FSA guidance on temperature control.

The regulations generally do not give specific temperatures other than a maximum of 8C for anything that needs chilling. While ripening some cheeses are so acid that bacteria cannot survive. These cheeses are ripened above 8C but must be reduced below 8C as soon as the acidity level falls.

- High-risk meats have lower limits, 4C for poultry and 2C for minced meats. This latter must be frozen below -18C, which is the only frozen temperature specified in the EC legislation.

- Milk must be below 8C, if it is to be collected every day, but below 6C, if collected on alternate days.
- Large game must be chilled to 7C, although *chilling may not be necessary where climatic conditions permit.*
- Other meat must be brought into the cutting room progressively to ensure that it does not exceed 7C or 3C for offal.

By not being specific in all cases the regulations put the onus on the food business operator to do his own risk assessment. After that "good practice" and the advice of food hygiene specialists should be sought. Some foods obviously do not need chilling (sugar or salt). Eggs are usually displayed ambient with a recommendation to store chilled. The reason for this is that eggs bought chilled would attract condensation in transit. Any dampness on the shell could transmit contamination to the inside.

Some foods, like fresh soft fruits, can be damaged by freezing because ice crystals destroy their structure. These may be preserved in inert gases, which inhibit the growth of bacteria or over-ripening.

Cut surfaces of meats are almost invariably contaminated which is why special precautions must be applied to minced meats. When cooking steaks the outside should be seared to reduce contamination. Where meat is cut there should be a supply of water above 82C for cleaning knives and equipment.

> Foods held hot should be above 63C. Reheated foods should be "piping hot", except in Scotland where they must reach 82C. These are legal requirements.

In summary, a *microbiologically unsafe food is one that contains levels of pathogens or their toxins, which could harm consumers when the food is eaten. All foods must be held under conditions conducive to maintaining their safety. The scientific assessment must clearly demonstrate that the microbiological safety of the food will not be compromised by storage and handling at the higher temperature* (FSA guidance document).This document runs to 17 pages and even then says it is only an outline.

Can you demonstrate that you keep all foodstuffs at the right temperatures? *Yes or No*	

Chapter 16: Training

"Train up a child in the way that he should go and, when he is old, he will not depart from it" - Proverbs.

The law on training is quite simple.

It then goes on to say that *those responsible for the development and maintenance of the procedure referred to in article 5 (this is the procedure based on the principles of HACCP) or for the operation of relevant guides have received adequate training in the application of the HACCP principles (see chapter 7).*

The FSA guide says that staff must be trained in food hygiene in a way that is appropriate for the work they do. Thus the law in Britain does not require any formal qualification, unlike in many other member states.

The question that should be put is "Do your staff appreciate the risks that their actions may involve?" If the answer to that question is doubtful, then some form of training may be needed to avoid the possibilities of cross-contamination, bad cleaning and other hazards. When selecting a trainer or training agency, make sure that they understand your business.

Two definitions that are used in Standards are useful:
- A skilled person is one who has relevant training, education and experience to enable him or her to perceive the risks and to avoid the hazards associated with the job.
- An instructed person is one who is adequately advised or supervised by a skilled person to enable him or her to perceive the risks and avoid the hazards, which the job may encounter.

The EC guide says that *training is an important tool to ensure effective application of good hygienic practice. It should be commensurate to the tasks of the staff in a particular food business and be appropriate for the work to be carried out.*
Training can be achieved in different ways. These include in-house training, the organisation of training courses, information campaigns from professional bodies or the competent authorities, guides to good practice etc. (EC Guide to 852/2004).

You must be able to demonstrate that both you and your staff have had the right instructions. Many of the larger retailers insist that their staff sign to say they have mastered each duty when it has been explained to them.

Typical Cases where training was lacking or needed

- "All my staff did the course at least ten years ago so why do they want more training?"
- The staff wore the hats provided but, in order to preserve their hair styles, most had loose hair protruding from the hat.
- Thinking that freezing killed bacteria.
- Trying to clean equipment with dirty cleaning cloths.

Can you demonstrate that you and all your staff have been trained correctly? *Yes or No*	
Is there a better training system I should use? *Yes or No*	

Appendices

Appendix 1

Other Rules

There are the UK rules. These are some:

- If you want to sell alcohol you must have a licence.
- If you want to sell hot food between 11pm and 5am you need a licence.
- If you want to sell from a van in the street you need a licence.
- If you have music playing you may need a licence.
- If you have live music you do need a licence.
- If your turnover (not profit) is more than the limit (£64,000) you will need to register for VAT.
- Ready to eat food and hot take-away foods are all subject to VAT.
- If you are self-employed you must register with the tax authority.
- You must have done a fire safety assessment.

Appendix 2

Abbreviations:

FSA — Food Standards Agency, responsible for consumer protection in UK.

EFSA — European Food Safety Authority.

EC — European Commission.

DG Sanco — Santé et protection consommateurs; Health and consumer protection Directorate General. Part of the EC.

HACCP — Hazard Analysis and Critical Control Points.

SFBB — Safer Food, Better Business. An FSA guide to food safety.

HSE — Health and Safety Executive, responsible for personal safety.

REACH — Registration, Evaluation and Authorisation of Chemicals. EC regulation.

CEN — Commission Européen Normalisation. Sets EU standards.

SI — Statutory Instrument. Device used by Government to pass laws without much debate in Parliament.

FBO — Food Business Operator. That is you.

SALSA — Safe And Local Supplier Accreditation.

OJ — The Official Journal of the EC where new laws are published.

MEP — Member of the European Parliament.

Appendix 3

Quotations in this Guide are in italics and are taken from the General Principles and Requirements of Food Law EC 178/2002, the Hygiene of Foodstuffs Regulation EC 852/2004, the Food of Animal Origin Regulation EC 853/2004 and their respective EC and UK guidance documents. All these have been brought into British law by a succession of Statutory Instruments.

References

Various websites have been quoted and referred to throughout this booklet, which will give access to further information and advice.

Food Solutions **www.food-solutions.org** has links to the two major pieces of legislation under "Useful links". The other regulation texts and official guidance are on the subscriber section under EC regulations. The explanatory articles listed alphabetically are in the subscriber section.

The Food Standards Agency is set up to ensure consumer protection, not to foster food businesses. However, it has references to most of the guides to good practice - **www.food.gov.uk**.

Safer Food, Better Business for England and Wales with its cousins CookSafe in Scotland and Safe Catering in Northern Ireland are useful introductions to hygiene practices and for developing your procedure based on the principles of HACCP. They can be obtained free from 0845 60 60 667.

Health and Safety issues are covered by the Health and Safety Executive **www.hse.gov.uk**.

This handbook has referred to the legal compliance, which may not be the ideal for your business. Good practice should take you further and ensure that your due diligence is justified. Guides to good practice have been issued by some sector specific organisations as well as by national bodies.

The Scores on Doors scheme awards compliance with regulations and good practice. See **www.scoresonthedoors.org.uk**.

Appendix 4

Glossary of Terms

Amendment	Slight change to a regulation after experience
Approval	Requirement for some businesses before trading. See chapter 6
Article	Mandatory clause in an EC Regulation
CEN	Commission Européen Normalisation. Standards body
CEN standard	European level document specifying benchmarks for equipment and procedures
Codex Alimentarius	International body that sets definitions for food
Competent authority	A nominated body in each member state that is responsible for enforcing the European food laws. In Britain it is the Food Standards Agency
Derogation	Compromise to allow some procedures to deviate from regulation.
DG Sanco	Health and Consumer Protection Directorate. Part of EC.
Directive	Decision by EC that has to be included in national law. Labelling as year/number/EC.
EC	The European Commission. The body that writes the rules.
Game	Animals other than those commonly farmed (cattle, bison, sheep, pigs, goats etc). For full definitions see Annex 1 of 853/2004.
Game Handler	Person who prepares game meat, after hunting, for placing on the market.
Home authority	Designated local enforcement body.
Home authority principle	The authority where the relevant decision making base of an enterprise is located.
Horizontal	Regulation that covers all businesses in a sector
Internal Market	Any sales within the European Union
ISO standard	International / Global standard, available from your professional advisor.

Local sales	Local is sales within your own county plus the greater of either neighbouring county or counties or 30 miles/50 Km from the boundary of your county.
Microbiological	To do with bacteria etc. that might cause spoilage or create hazards
Micro criteria	The limits of bacteria etc. allowed in foods.
Potable water	Water of drinkable quality.
Pre-packed	Goods in final sales permanent packaging (see chapter 14.1).
QUID	Quantitative Ingredient Declaration, on labels where an ingredient is added.
Recital	Clause in the introduction to a regulation to explain rationale.
Regulation	Decision by EC that is binding in entirety on all member states. Labelling as number/year/EC. Note, these numbers are not unique to food, but are sequential for all decisions.
Renderers	Firms that heat-treat meat and bones.
SALSA	Safe and Local Supplier Approval. Scheme to verify that a food business has reached certain high levels of management practice.
Scores on Doors	Scheme where businesses are marked after each local inspection.
Toxin	Poison that is resistant to cooking, created by bacteria etc. in food.
Transitional	Temporary regulations to allow business time to comply
Use by	The only mandatory durability date (see chapter 13.2).
Vertical	Regulation that is specific to a class of foods (Honey, Jam etc) or a process
Zoonoses	Diseases caught by humans from animals e.g. TB, rabies etc.

Index

What now?

For food hygiene and safety you need a plan to correct things which may cause hazards. To find out how well you are complying with the law, you can use our checklist. This takes you through all the areas the food regulations cover and gives you an indication of how well you are complying. Having completed the checklist you can identify your strengths and weaknesses and take steps to improve weak areas well before your next inspection.

Food Solutions can help keep you up to date with the regulations, provide training aids and record sheets which help you demonstrate you are doing what you say you are doing. There is also a short version of this book which gives you the basics and is backed up with the full online handbook which is updated as necessary.

For details of the Food Solutions products and how to obtain them see:

www.food-solutions.org

Useful Numbers	
My local Environmental Health Office	
My local Trading Standards Office	

The table below details the main areas you need to consider in your food business. It also tells you where to find information and advice.

Subject	Information	Advice
Registration	FS handbook	EHO
Traceability	FS handbook	TSO
Personal Hygiene	SFBB DVD	
Premises	FS handbook	EHO
Equipment	FS handbook	suppliers
Pests	SFBB	BPCA
Labels	FS handbook	FSA/TSO
Allergens	FS handbook	FSA
Contamination	SFBB	
Cleaning	SFBB	suppliers
Temperature	FS handbook	SFBB
Training	FS handbook	EHO

Abbreviations.

FS= Food Solutions;

EHO= Environmental Health Office;

TSO = Trading Standards Office;

SFBB = Safer Food, Better Business pack available from your EHO;

BPCA = British Pest Control Association www.bpca.org.uk

Food Solutions Products

For full details visit www.food-solutions.org

Product	Format	Aim
Handbook - Understanding Food Hygiene and Safety Regulations	Hard copy	Tells the reader why the laws are there, who has to comply and what they say you have to do to comply
Quick Reference Handbook – Food and Safety Regulations made easy	Hard copy	Outlining the essential parts of food regulations
Handbook – Understanding Food Hygiene and Safety Regulations on line version.	Electronic	Used in conjunction with the Quick Reference Handbook. The E-Handbook keeps readers up to date with Food Regulations.
Check-list - Understanding Food Hygiene and Safety Regulations	Hard copy Electronic	Completing this initial review check-list will tell FBOs where they are weak as far as compliance is concerned.
HACCP Check-list	Hard copy Electronic	Lists the areas that need addressing to demonstrate HACCP within the food business
Advice Modules	Electronic	Aimed at the owners of food businesses. To inform them about a wide range of topics
Training Modules	Electronic	Aimed at supervisors and staff within the business. Offering documented proof that training has taken place on a wide range of subjects.
Management Records	Electronic	Enables the FBO to record events (e.g. cleaning schedules).This helps demonstrate due diligence to the authorities.
Update information-quarterly magazine	Hard copy	Keeps FBOs up to date with proposals and changes to Food Regulations.
Update information - Bulletins	Electronic	Regular updates on issues that could affect FBOs